A Preface to Literacy

LESSON XXIII.

ha	cold	kept	found	cru-el
try	rain	died	James	a-way
fire	none	took	young	be-gan
Ann	road	some	would	look-ed
who	poor	chirp	ground	wick-ed
flew	from	wants	thought	naught-y

THE YOUNG BIRD.

JAMES BLAND found a poor young bird on the cold ground. It was all wet, for there had been rain that day.

"Ha!" said he, "I will have a fine pet, now." So James took it home. He met his sister Ann at the door.

A Preface to Literacy

An Inquiry into Pedagogy, Practice, and Progress

Myron C. Tuman

The University of Alabama Press
Tuscaloosa and London

Publication of this book was made
possible, in part, by financial assistance
from the Andrew W. Mellon Foundation
and the American Council of Learned Societies.

Library of Congress Cataloging-in-Publication Data

Tuman, Myron C., 1946–
  A preface to literacy.

  Bibliography: p.
  Includes index.
  1. Literacy—United States. 2. English language—
Study and teaching—United States. 3. Tuman, Myron C.,
1946–   . 4. Teachers—United States—Biography.
I. Title.
LC151.T86 1987   428   85-29010
ISBN 0-8173-0307-3

British Library Cataloguing-in-Publication Data
is available.

To Walter and Jeremy
my father and son

Contents

Acknowledgments

This book, very much a theoretical discussion of the nature of literacy, is also a spiritual autobiography, reflecting in countless ways my life as an educator, from seventh-grade English teacher to director of graduate seminars in basic writing. My career, like that of the naive, young protagonist of any bildungsroman, has been aided immeasurably by a series of benefactors who, in the spirit of David Copperfield's Aunt Betsey, have surfaced at the most opportune moments with offers of sound judgment and generous, kindly support. First, deserving of special notice, is the National Endowment for the Humanities, which provided me with a year-long residency at the University of Southern California, during which I undertook the initial research for this book. I must also thank a series of chairpersons, beginning with Gail Little at the Louise S. McGehee School, followed by Raeburn Miller at the University of New Orleans and Rudy Almasy at West Virginia University, and ending with Claudia Johnson at The University of Alabama, who all expressed confidence in the value of my work. Third are colleagues, from John Clifford and Don Lazere at the seminar of the National Endowment for the Humanities and Winston Fuller and Barry Ward at West Virginia University to Jim Raymond at The University of Alabama, who at various times greeted my critical excesses with bemused sympathy and sound analysis. I am also grateful to the clerical and administrative staff at West Virginia University, especially Jackie Seymour and my own secretary, Linda McCord, who, by their competence and unflagging good cheer, provided a harried administrator at a writing center with many welcome and unexpected opportunities to work on this manuscript.

Finally, I thank my wife, Ginny, who, while I was preoccupied thinking and writing about education, daily faced the more demanding and more important task of teaching children—the countless fifth and sixth graders in the public schools of Louisiana and West Virginia and Alabama, as well as our own three children—just what it means to read and write. Any truth that this book speaks about literacy is the product of a collective wisdom, more theoretical on my part, more practical on hers, born of this loving relationship.

Material from chapters 2 and 3 appeared in considerably different form in two articles in *College English:* "Words, Tools, and Technology" (1983) and "From Astor Place to Kenyon Road" (1986).

Acknowledgments

A Preface to Literacy

Introduction
Literacy Today

What is literacy, and why is it valuable? The answers to these two questions seem fairly straightforward as long as one proceeds from the premise that literacy reflects powers inherent in language generally. The craft of the storywriter, accordingly, can be explained in terms of the craft of the storyteller. After all, both expertly use language to describe what has happened elsewhere, taking what British educator James Britton ([1970] 1972) calls the role of spectator (rather than the role of participant). When we use language as a participant, Britton explains, we are trying to accomplish something practical. We are engaging in some activity of direct and immediate concern to us; we are, in a sense, actually playing the game. At other times, however, we are spectators, content to observe and comment upon what has happened or might happen elsewhere. At such times we reminisce or predict, perhaps even to ourselves; we become storytellers, creating in language for those in our presence another, less immediate world.

The story being narrated must be given life in language, even if it is already well known to the listeners, since the pleasure it provides resides largely in our experience of the power of the teller and, indirectly, in our awareness of the power of language itself to transcend the limits both of everyday verbal interaction and, more important, of the social bonds that ordinarily govern our conduct. When we are engrossed in a tale, in either its spoken or its written form, we enter a world in which human action (here imaginatively represented in language) is not as fully restrained as it is by the customary real-world demands of etiquette, decorum, and propriety. The transportive power of such narratives, in other words, seems to be largely a function of the liberating potential of language in general and not of writing per se. Thus it seems possible to define literacy largely, if not entirely, as the technical skill necessary to transform the oral version into its written form. We therefore find our task quickly changing shape before us, shifting from a very general discussion of the value of language to a quite limited discussion of the nature of transcription and sometimes alternating between the two. In either case there seems to be little of interest to be said about the concept itself. We may thus be tempted to begin a quest for a more specific understanding, as the protagonist of Bunyan's *Pilgrim's Progress* first approached the open road before him, with little sense of the need to prepare for a long journey.

Matters are not as simple as they seem, however. How, we might ask, can we invoke *Pilgrim's Progress*, given that Bunyan composed his tale while he was a prisoner in the Bedford jail over three centuries ago? Although spoken and written narratives show important similarities, the actual products of the storyteller and the storywriter are not identical. Whatever stories Bunyan might have told his fellow prisoners, for example, are largely lost without written records. Although a memorable story, by definition, stays with us long after its first hearing, and thereby achieves some of the independence of a text as our experience with it deepens and shifts, the initial encounter itself is necessarily embedded in a specific moment of social interaction, a time when our relationship to the storyteller mediates our relationship with the story. While both the storyteller and storywriter transport us into another world—such is the creative power of their language—the future of the spoken tale depends largely upon its teller's ability to strike a receptive chord within a specific audience. The spoken story may form a link in a long chain of human interaction, perhaps uniting people in a way that its written counterpart may not, but such contact also requires that the chain remain unbroken.

In contrast, the permanence of the written version depends less upon a set of historical links to a specific event in the past. As readers of *Pilgrim's Progress*, we need no personal contact with Bunyan or with a chain of events going back to his imprisonment. In this regard our relationship with Bunyan resembles our relationship with writers, from Boethius to Martin Luther King, who have addressed us from a literal prison or have worked in physical isolation—from Anne Frank, keeping her diary in a warehouse in Amsterdam, to Proust, composing novels in a cork-lined room in Paris, to Emily Dickinson, sequestered in a house in Amherst, to Machiavelli, writing about political power while exiled at his villa outside Florence. Indeed, this image of the writer's imprisonment, whether literal or figurative, suggests one key element in storywriting and hence in literacy that need not be present in storytelling—namely the impulse to communicate with people who are not within speaking distance.

The ability to address individuals not present may be facilitated by, without entirely depending upon, the most visible component of literacy: the ability to transcribe words. Whether or not something is written, therefore, may be only evidence, and not conclusive proof, that one key element of literacy is present—an active desire to address an audience that may or may not be distant, or a readership. "Whereas we can apply the term 'literacy' to an individual," writes classicist Eric Havelock, "its operant meaning derives from the fact his literacy is shared by a given number of people, all of whom are readers" (1976:20). In exercising literacy, Havelock adds, people without prior contact place themselves "in automatic communication with each other." Indeed, we find this ability to address distant readers such a compelling component of literacy that we customarily call someone a "writer" even when it is apparent, as with Milton's composition of *Paradise Lost*, that the text was dictated. Similarly, we regularly feel that certain creative feats of oral language, such as the recounting and reworking of traditional stories in formulaic narrative verse, require some of the very skills that are suggested by the term "literate," especially when these skills are compared with the limited ones necessary for someone to parrot a string of loosely related words written on the page. Literacy thus seems concerned as much with the ability to use language in certain engaging and unusual ways as with specific transcription skills.

Such reasoning leads to a seeming paradox: we can call someone "literate" without knowing for certain whether he or she is able to read and write. Although we may often seem to do so, such common usage does not make it at all obvious what the designation really means. Just what skill or set of skills is required, if not the ability to read and write? Is there possibly more than one kind of literacy, and, if so, how are the different sorts related? Is there any connection between the varieties frequently discussed today—between "computer" and "television" and "mathematical" and "cultural" literacy? People certainly do need to learn—and educators must consequently identify and teach—many different skills for use in many different situations. The reason for using "literacy" to denote the mastery of such skills is unclear, however, inasmuch as some of them have no obvious connection with reading and writing. And unless we have a greater sense of the meaning of literacy, how can we be certain about its value? If successful functioning in society is the primary goal of education, then perhaps we mean that we value the capability (or capabilities)—entailing a myriad skills—of doing what we want to do with our lives, that is, we prize all the compound forms of literacy, especially the one known as "functional literacy."

In this regard, we must recognize something highly unusual about present discussions of this key word, both in the general population and in academic circles. While educators, psychologists, and anthropologists regularly discuss "literacy" in terms of minimal reading and writing skills, literary theorists, composition specialists, and hermeneutic philosophers rarely use the term when discussing the problems entailed in generating and comprehending texts. As a consequence, it is not obvious what questions are being raised when a historian discusses literacy in terms of the percentage of people able to sign their name rather than merely make an X on some official record such as a marriage certificate. While we obviously do need to know when various groups of people achieved such minimal transcription skills, we should also more clearly know how such a historical discussion informs current debate about the "literacy crisis" in a postindustrial society that regularly mandates formal schooling for all children until age sixteen and readily gives practically the whole post–high school population access to some form of higher education.

One problem we face today is that any reference to the "literacy crisis" is likely to mean either that many older Americans have never learned to read and write at all or that high school and college students do not read and write very well despite our educational effort—or both at different moments. Consequently, while there is undoubtedly widespread public concern about the current state of affairs, we can never be certain which set of problems is in question. It is hardly surprising in this context that Terrel Bell, a recent secretary of education, writing in a syndicated Sunday supplement, ended a brief note on the declining state of "literacy" and education generally with a plea that "our grandchildren may conquer illiteracy" (1982:2), suggesting that our most pressing problems in becoming a fully literate society will be overcome when everyone has achieved a certain minimal level of performance.

Many people in the United States still never learn to read and write, as shown in Jonathan Kozol's recent study, *Illiterate America;* many others, however, drift through years of formal schooling, some of it in colleges and universities, attaining only the most superficial skills. The "literacy crisis" of the unschooled and the uneducated, which Kozol discusses, is both real and threatening but perhaps no more so than the "literacy crisis" of the "educated." And this latter crisis, it is important to realize, may have little to do with whether reading and writing skills are declining or even improving. The real threat to literacy today, in other words, may reflect less the disappearance of books themselves or our inability to read the words in them than our increasing inability to interpret the worlds they evoke. What good would it do us to read all the words in a book if at the same time we were unable to see the world embodied in it as both different from our own and a commentary on it? What if we could "read" a book but were still unable to see it as exposing us to, in Wittgenstein's words, a new "form of life"? In this regard the real threat to literacy may be that the verbal meanings in written texts seem increasingly irrelevant to more and more people who have the coding skill, but not the compelling motivation, to interpret them.

We read texts not in isolation but, as literary critic Stanley Fish (1980) notes, only as part of "interpretive communities." What if such a community were to change radically, so that the goal of literacy changed as well? Would we then start becoming more or less literate? More to the point, do we have any assurance that we would collectively be able to notice the change? After all, the recognition of certain physical components of literacy, such as reading aloud and taking dictation, requires only marginal skills very different from those needed to iden-tify successful acts of verbal creation and comprehension. Who are we to judge the quality of other people's reading and writing, anyway, and for what purpose? Indeed, to the extent that one act of literacy can be recognized only through another, we might suppose that a collective decline in our ability to create and comprehend texts would be accom-panied by a general shift toward a mechanical definition. Educator John Bormuth, for example, speaks of "the actual exchange of information that takes place when a person reads a material" (1978:19). Bormuth contends that economic forces govern the amount and the quality of our information processing and that therefore we can "generally count on ordinary economic forces to ensure that literacy is being produced efficiently" (p. 38).

As long as we can provide enough technicians so that information flows freely and steadily between data banks—and is there any doubt of our ability to do so, given the extent of our educational effort?—we should be able to claim continued high levels of "literacy." Levels of literacy, therefore, if we understand the term to mean certain mechan-ical aspects of verbal transmission, could well appear to be increasing at the very moment when we as a people are becoming less literate. Meanwhile, fewer and fewer people may be interested in whether "literate" technicians, or any of the rest of us, can do more than store and retrieve information. How important will it be, for instance, whether we are still able to voice our thoughts and to understand those of others? Even more important, how likely are we to see any point in such an activity?

Contemporary discussions of literacy are often, in the words of historian Harvey Graff, "confused and ambiguous," a surprising finding, given the pervasive belief that the introduction of writing brings an increase in the precision of verbal analysis. "Vagueness pervades vir-tually all efforts to discern the meaning of literacy," Graff adds. "More-over, there is surprisingly little agreement on or special evidence for the benefits of literacy, whether socially or individually, economically or culturally" (1979:3). All the polemics and national debate about literacy notwithstanding, we still have few answers to the two basic questions addressed in the following chapters: just what is literacy, and why is it valuable? Indeed, it is possible to contend that, amid all the public discussion of computer, cultural, and functional literacies, there is at the present time little discussion, public or academic, of *literate* literacy—of literacy construed as the skills necessary to comprehend and create the kinds of texts traditionally communicated in written form.

It is one thing to be able to explain many aspects of cultural life in terms of the rise and fall of literacy; it is a different matter to consider the rise and fall of literacy in terms of changes in culture. For the most part, scholars have been too interested in literacy as a cause of complex historical changes and thus too willing to consider it as something fairly straightforward and obvious. We have used the concept to analyze many important issues about cultural change, about industrialism and economic growth, and about education and social policy, but we have addressed few questions about the special nature and value of reading and writing. As a result we have reached certain fundamental insights regarding cultural life but comparatively little understanding of the special nature and value of written discourse. If this trend is to be reversed, we must begin thinking self-reflexively, by raising the most basic questions about the very discourse of which literacy speaks and of which, being written and read about, it is.

Part One
Pedagogy

1

The Nature of Literacy

I. Until recently, the Library of Congress subject heading for literacy read "LITERACY—see ILLITERACY." Why, for most of this century, was there no separate listing for "LITERACY?" Did the catalogers assume that it was possible to define the one as the opposite of the other? There is a general consensus, built upon common usage, about what it means to be literate: "That person is literate who, in a language he speaks, can read with understanding anything he would have understood if it had been spoken to him; and can write, so that it can be read, anything that he can say" (Gudschinsky 1976:3).

Such a definition comprises what can be called an unproblematic model—unproblematic, not in ridding the study of literacy of all problems (many areas of contention remain), but in preventing questions about the nature of literacy from being placed at the very center of deliberations. The basic issues that the unproblematic model raises are practical rather than philosophical: how can we set measurable standards of proficiency in coding so that we can determine the number of illiterates and so that educational programs can be designed, implemented, and evaluated? Should people be able to encode as well as decode, and how do we classify those who can perform one but not the other? Should related abilities such as listening be included? How should we treat essential nonverbal skills such as the ability to add and subtract?

Educators and social planners must answer such questions in the regular conduct of their work. Although they do often grapple with the problem of a definition, their concern is less with the nature of literacy itself and more with the bureaucratic requirements for collecting data and formulating educational policy. With the unproblematic model, as in a territorial dispute, the subject of contention is the definition of boundaries—solely the problem of specifying the exact parameters, or the acceptable limits, of the coding skills required. The basic issues, as such, are the same whether one is considering "literacy" or "illiteracy"—the difference is merely that one term refers to the presence of certain skills and the other term to their absence.

The dominance of the unproblematic model derives in part from its reliance on skills that are readily observable and measurable. Educators seem especially attracted to a model of literacy that has the power to reduce theoretical issues to familiar patterns of behavior. Questions about the complex nature of reading, for example, can be reformulated as dealing with the extent and degree of decoding:

Learning to read . . . is *not* a process of learning new or other language signals than those the child has already learned. The language signals are all the same. The difference lies in the medium through which the physical stimuli make contact with his nervous system. In "talk" the physical stimuli of the language signals make their contact by means of sound waves received by the ear. In reading, the physical stimuli of the same language signals consist of graphic shapes that make their contact with his nervous system through light waves received by the eye. The process of learning to read is the process of transfer from the auditory signs for language signals which the child has already learned, to the new visual signs for the same signals. [Fries 1962:xv]

Pushed to its logical limit, such thinking equates the principal source of reading problems with the organ of physical contact: "The 'reading problem,' as we know it," write two reading specialists, "would not exist if, in dealing with language, all children could do as well by eye as they do by ear" (Jenkins and Lieberman 1972:1). This same reliance upon an unproblematic model of literacy is reflected in a report that a Harvard committee on composition wrote some eighty years ago. The report observes that it is "little less than absurd to suggest that any human who can be taught to talk cannot likewise be taught to compose. Writing is merely the habit of talking with the pen instead of the tongue" (quoted in Kroll and Vann 1981:vii).

The unproblematic model supports the belief that our most fundamental problems involving reading and literacy would disappear were all students equipped with cassette recorders. Then they could merely record rather than encode what they wanted to say and could listen to rather than decode what they wanted to understand. Everything would work fine as long as people sought to express or understand essentially the same message that they ordinarily communicated orally.

Any new technology can eliminate various pedagogic difficulties by substituting an easier technology for a more difficult one in the same way that telephone calls have replaced some functions of letter writing. It remains in doubt, however, whether various new technological forms of oral communication and storage can totally replace writing or whether they merely reduce the need for it. After all, we do continue to write letters, even though we could more readily make a telephone call.

Pen and paper obviously do provide a permanent record, useful in legal contracts, but they may also allow us to generate more detailed messages than we could readily communicate by telephone, for example, statements that rely on the often subtle distinction between restrictive and nonrestrictive modifiers. Thus, while speech is often a more efficient way of sharing information, writing sometimes more efficiently specifies complex subordinate relationships. One corollary is that reading comprehension is often higher than listening comprehension for proficient decoders confronted with complex material (Sticht et al. 1974). To the extent that it is desirable to identify subtle hierarchical relationships in verbal communication, reading and writing may embody technologies that cannot so readily be replaced by oral recording. Reading and writing, quite simply, may often be hard, regardless of how well we can code oral language, because the content of the communication is difficult or unfamiliar.

With the unproblematic model of literacy, we gain categorical certainty at the cost of any real understanding of the nature of literacy itself. This model, for instance, offers no clue as to why we value reading and writing so highly. Is it because writers possess some magical coding ability that enables them to make permanent what is otherwise transitory? Is our collective concern with literacy only a fetish for a certain kind of verbal technology? The unproblematic model, furthermore, does little to explain why reading is so difficult for many people (as evidenced by the fact that some students have still not mastered the skill in high school, while others are fairly competent before they enter first grade). And why does writing predictably present even greater difficulties? How, then, do we explain the fact that very many people with reading and writing skills have only limited use for them? What cultural conditions lead people actually to see the value in using these skills in verbal acts of creation and comprehension? Only after we have answered such questions can we decide whether the basic challenge of literacy education is to teach students to code language, or to motivate them to create and comprehend texts. If the latter, then whether or not students can process language as well by eye as by ear may bear only indirectly on the current concerns about literacy.

The unproblematic model, while an efficient means of allowing educators and others to survey populations and to evaluate programs, is not necessarily an accurate phenomenological account of literacy itself. The practical reasons for defining reading and writing as coded forms of speech may bear little relationship to our actual experiences with written language. It is one thing to suggest that a person who can encode language can write and is thus to be considered literate; it is a very different matter to suggest that writing *is* encoded speech. Indeed, it may make more sense to assume that the world of speech and the world of writing are fundamentally different, noting that, while people who are literate may be able to write what they say and say what they write, what people actually say and actually write rarely coincide.

People seldom write down what they say or speak what they have written down, and on the occasions when the two categories do overlap, as when we may write down notes in preparing to give a speech, the words often have a peculiar, hybrid quality. Indeed, there is so little reason to write down what we *ordinarily* say that we have few transcriptions of anything even resembling conversations before the rise of realistic fiction in the last two and a half centuries—and it is doubtful whether we have any actual transcriptions of ordinary conversations before the recent efforts of sociolinguists and discourse analysts.

Clearly, writing has as one technological advantage that it can substitute for memory, not for speech; accordingly, if one were forced to choose between the two positions, it would be more accurate phenomenologically to define literacy in terms of the *skills necessary to encode what one would otherwise memorize.* Such a change, as simple as it may appear, nevertheless does shift the focus of our attention from the evanescent world of speech to a consideration of the kinds of things that people have historically stored in written form.

It is not especially difficult to classify the things that were written down for most of recorded history. The 150,000 extant Assyrian texts, for example, fall into two major categories (Wiseman 1962:22). Three-quarters of all the Assyrian texts consist of legal documents—deeds of sale, promissory notes, wills, inventories, and other such records, some of which, like inventories of ordinary commercial merchandise, do not ordinarily take a spoken form. The precursors of the list and similar records were not spoken but concrete—knots, tallies, and other mnemonic aids. Moreover, the creation of an inventory may require the ability not only to encode speech but also to organize seemingly diverse materials into coherent classes, a skill that manifests no obvious connection with speech and that conceivably has a very different origin.

The other group of Assyrian texts, representing only a fourth of the total, includes transcribed poems, prayers, legends, and myths—items that, unlike the legal documents, were transposed from an oral form to a written one. Here too, however, it is misleading to say that individuals transformed into writing verbal messages that were ordinarily spoken. Writing was being used, not to transcribe speech, but to confer greater permanence on material that was already preserved mnemonically—material that was regularly recited and was thus preserved, albeit in a nontextual and hence more fluid format, long before it was ever written down. The *Rig Vedas,* Goody notes, were not transcribed until nearly a thousand years after writing had been introduced into the area; this finding supports the folk belief that books not memorized were comparable to wealth in the possession of others (1968a:12–13). Thus where such religious, nonlegal texts are concerned, we must really distinguish not between what is spoken and what is written but between what is spoken in the course of ordinary social dealings (and remains in oral form even after a society becomes fully literate) and what is first recited according to some complex, perhaps ritualistic pattern of oral mnemonics and is only later written down.

The transcriptions of oral material are also unusual in that the scribe responsible for encoding the material does not necessarily bear any relation to the person, or the collective cultural experience, responsible for generating it. "Writing," under such conditions, means only the ability to copy manuscripts with speed, accuracy, and elegance, while "reading" means only the ability to recite something using the text as a mnemonic aid. Writing in such a world is a branch of calligraphy, and reading is akin to the practice, still current in some religions, of praying by reciting transliterated versions of prayers. "Readers" may thus know only the appropriateness of what they recite, not its meaning. Meanwhile the content of the transcription may be guarded by an elite or priesthood whose standing is enhanced by the strangeness of the texts and by the responsibility of rendering them in writing. Indeed, many peoples throughout history read and wrote in a language other than the one they ordinarily spoke; in such situations not only was the written language not used in ordinary speaking situations, but the spoken language sometimes had no written form whatever. The most accomplished readers and writers, moreover, might not even have been "literate" according to the unproblematic model—that is, they might not even have been able to encode and decode the statements they ordinarily heard and uttered.

As linguist George Dillon (1981) notes, many of the writing problems of college freshmen can be traced back to their unfamiliarity with the conventions of the expository essay and in turn to their unfamiliarity with texts generally. Although the philosophy teacher in Molière's *Bourgeois Gentleman* notes that we speak prose all our lives, students who have done little reading or writing in the past regularly have trouble producing the extended, developed prose that forms the basis of academic discourse. Indeed, the difficulties that students often experience in mastering prose reveal something important about the phenomenological nature of writing itself.

If, for example, we rely on complex patterns of subordination more in writing than in speaking, then we have an added problem in mastering prose: how to give discourse that is structurally and functionally different from speech the same easy spontaneous quality. In writing, we regularly want to make statements that we would hardly ever speak aloud sound as if we had just said them. Prose, in other words, is writing masquerading as speech; teachers should therefore advise students to write as naturally as they speak only when the students are already capable of speaking naturally and at length before groups of strangers. (It is easy to write as naturally as one talks if one is an adroit public lecturer.) Few students, however, master the technique of speaking so as to command the attention of strangers for prolonged periods of time, and while some writing problems do reflect students' failure to master coding skills, many others entail problems of point of view, audience, diction, syntax, organization, and development that may affect their efforts at writing or speechmaking equally.

A report of the National Assessment of Educational Progress, *Reading, Thinking, and Writing*, clarifies this issue by recording the changes in the ability of various age groups over the last decade to form judgments in writing on the basis of material they have read. It concludes in part that "very few students at any age *explained* their initial ideas and judgments through reference either to the text or to their own feelings and opinions" (p. 1). This problem resulted from an inability to use evidence to support opinion: "The overwhelming majority of students lacked strategies for analyzing or evaluating in the interest of deepening their understanding of what they read" (1981:1). Comparing this study with one conducted a decade earlier, the researchers conclude, "One way of characterizing the change during the seventies is to say that 17-year-olds' papers became more like 13-year-olds' papers" (p. 3).

The unproblematic model of literacy misleads us into believing that the change, although perhaps important for a number of reasons, says little about literacy itself. Yet the NAEP study was more likely concerned with the very essence of literacy and the unproblematic model of literacy with only the peripheral issue of transcription. Indeed, the principal goal of the unproblematic model is not to define literacy itself but only to specify the point at which a person can no longer be considered illiterate. The widespread acceptance of this model merely reflects the important, although largely utilitarian, goals of educators concerned with teaching basic skills to large numbers of people who are generally cut off from both political power and material wealth.

Educators may continually refer to the problems of "literacy," but their real interests are often elsewhere. "Rather than an end in itself," notes a 1976 UNESCO policy statement, "literacy should be regarded as a way of preparing man for a social, civic and economic life that goes far beyond the limits of rudimentary literacy training consisting merely in the teaching of reading and writing" (1976:10). This statement reflects the vital concern of most "literacy" educators throughout history and throughout much of the world today with helping masses of people cross a line, for years arbitrarily defined as a fifth-grade reading level, to become literate. Few can question the practical importance of this effort, just as few can expect such educators to be overly concerned with the theoretical issues at stake in our regular use of "literate" to describe both an average fifth grader and a world-renowned author. To consider this latter issue is to raise questions about the nature of literacy itself, and to do so we need an entirely different model of literacy, strictly based neither on coding skills nor on any formal features of language.

To understand *literacy* rather than the *boundaries of illiteracy*, we must stop relying on the most obvious and superficial differences between spoken and written statements. While literacy does entail reading and writing, it cannot be defined simply in terms of transcription and decoding skills. It is always tempting to see the formal differences between what is spoken and what is written, which are both necessary and readily observable, as the *essential* differences and hence as a ready means of distinguishing those individuals who are literate from those who are not. Graphemes and morphemes are different, as Joseph Vachek asserts in his study *Written Language*, but any effort to locate the essence of literacy in the differences between them is doomed to triviality, since the great accomplishment of phonetic transcription is that it permits people to shift freely and easily between speech and writing. Almost anything that we can say we can write, and almost anything written can be spoken—yet this fact only obscures the far more important point that much of what is written down never existed as speech and little of what is spoken is ever written down (nor would it be likely to serve any purpose were it so rendered). Written and oral communications can easily overlap; the key to understanding literacy, however, lies in recognizing the inherently different functions each mode of discourse more readily fills.

II. The central problem is simply stated. How can we define literate communication with reference to meaning rather than to coding skills? How can we demonstrate, in other words, that certain verbal messages belong to the world of writing and others to the world of speech on the basis of their content rather than the mode of transmission? Up to a point, this problem is not as difficult as it may appear. Consider, for example, the conundrum of a bottle washed up on shore containing the following cryptic message: "Meet me here at this time tomorrow with a stick this long." Many of the key words and phrases in this sentence—"me," "here," "this time," "tomorrow," and "this long"—are deictic terms, not meaningful in themselves, but signaling meanings that must be shared by the sender and the receiver of the message and that stand apart from the message itself. Writing often utilizes such deictic terms, but to do so it must provide a larger frame of reference, either elsewhere in the text or possibly in some preexisting, shared understanding between writer and reader. It is doubtful, however, whether any degree of intimacy between parties could supply a suitable referent for "this time." If a signature, an address, and the time and date were given, the reader could supply much of the missing information, but the phrase "this long" seems to have a visual referent, conceivably one that could be conveyed graphically by a scaled diagram or, perhaps more likely, by hands held a certain distance apart during a face-to-face encounter.

Since the message lacks a diagram, an address, and a signature, the circumstances in which it might have been written in the first place are hard to imagine. There are possible scenarios (for example, the receiver of the message is in a position to see the sender but not to hear or to understand the sender's spoken words), but any such explanation takes us far from the world of ordinary communication. There is simply no ordinary situation for writing such a text; it was merely invented by a linguist to argue a point about the role of context and shared understanding in communication. Even though few people would have any difficulty either reading or writing this message, it cannot readily serve as a piece of writing. Its meaning is almost totally embedded in context. It belongs to the world of texts only because it is in written form; as a piece of actual discourse, it belongs to the world of speech.

Other isolated sentences can just as easily be classified as pieces of writing: for example, the sentence "Nowadays people know the price of everything, and the value of nothing." This sentence, unlike the first, can be placed in a specific, albeit fictitious, context. Here Lord Henry is interrupting a conversation between his wife and Dorian Gray in Oscar Wilde's novel. As with the first sentence, there is a deictic reference, "nowadays," referring to the late Victorian setting of the story, thus making Lord Henry's remark a nominal comment on his preceding sentence, "I went to look after a piece of old brocade in Waldour Street, and had to bargain for hours for it."

Yet the satiric thrust of the remark about price and value goes far beyond its reference to Waldour Street in late nineteenth-century London. It is in turn part of Lord Henry's effort to loosen Basil Hallyard's conventional, moralizing grip on Wilde's youthful protagonist and part of Wilde's own efforts to shatter late Victorian social illusions and indeed such illusions generally. For this reason, we feel no constraint in the deictic reference to "nowadays"; as an aphorism, the sentence transcends any one context. While the actual sentence exists today in written form, although as part of an imagined dialogue in a novel, it belongs to the world of texts to the extent that in some vital sense it transcends its immediate context.

Conversely, the sentence about Waldour Street, immediately preceding the aphorism, like the sentence about the bottle, is embedded in its context, with the result that its full meaning depends upon someone's knowledge of the specific situation in which it was uttered. Specifying the situation would make either sentence more meaningful to a wider audience but would likely not increase either one's value, since that value resides in the narrow context of its actual utterance. It is thus possible to contend that much of what is spoken would serve no purpose were it written down. With the aphorism, however, the situation is quite different. Here too, context may be important—our knowledge of it can add to or, given the possibility of irony, even radically alter our understanding of what is being asserted. This shift in meaning, however, is largely comprehensible and intrinsically valuable apart from its location in this specific situation. Indeed, the sentence is an aphorism precisely because it does objectify a meaning independent of context. With the other two sentences, we must first grasp the context in order to grasp the utterance; here we must understand the utterance in order to understand the context.

The difference between Wilde's aphorism and the message in the bottle raises the most fundamental issue about the nature of literacy— the relationship between meaning and context. As compared with ordinary conversation, the communicative power of literate texts (whether in written or spoken form) requires of both readers and writers the negotiation of meanings that rise above, transcend, or in some sense cross expectations based upon ordinary social experience. In establishing meaning independent of expectations, literate texts fundamentally extend the power of metaphor to redescribe the world. While language can and often does affirm our interpretation of the world, in metaphor it exhibits the power to unfold before us a new, unexpected way of understanding by forcing us to recognize that a literal reading of an assertion will not work, that a given verbal account of the world is inapplicable, given a specific context, and therefore that, if the account is to be taken seriously and not dismissed, we must interpret it in a new or "figurative" way.

"The strategy of discourse implied in metaphorical language," writes philosopher Paul Ricoeur, "is neither to improve communication nor to insure univocity in argumentation, but to shatter and to increase our sense of reality by shattering and increasing our language. The strategy of metaphor is heuristic fiction for the sake of redescribing reality" (1978:132–33). It is a lie serving to extend truth, and the ontological status of the metaphor as a lie is most crucial for understanding the nature of literacy. While the interpretation of a metaphor will eventually lead us back to a familiar context, its digressive nature—its ability to force us to take seriously what we must first recognize as an inappropriate and ultimately unintended meaning—is most important. In trying to understand the phenomenological basis for literacy, in other words, our principal concern is not with the indirect truth of a metaphor (what is sometimes called its *tenor*) but with its direct, literal referent (the aspect of the metaphor that is sometimes called its *vehicle*), which is a literal lie.

These issues can be clarified by considering the working of an actual metaphor, here one suggested by philosopher John Searle (1979). In Disraeli's claim that he had "climbed to the top of the greasy pole," Searle notes a literal meaning (what he calls "sentence meaning," namely the performance of some athletic stunt) and a figurative meaning (what Searle calls "utterance meaning," namely the difficulties of attaining high political office). We understand the metaphorical assertion only if we understand the utterance meaning—Disraeli is talking about politics, not athletics—but we can understand the utterance meaning only by first grasping the sentence meaning and then realizing its inappropriateness (it is unlikely, given our historical knowledge of Disraeli, that the speaker would be making such an assertion). We must recognize, in other words, both that the utterance refers to certain difficulties involved in climbing a pole and that Disraeli would be unlikely to have meant the actual feat. We interpret the utterance as a metaphor precisely because its sentence meaning is unsupported by context and, equally important, because it is unintended.

The sentence meaning Ricoeur calls a lie, since it necessarily conflicts with the dictates of context and intention (if it did not, it would not be a metaphor); yet in order to reject the sentence meaning for some figurative reading, we must first understand it. The power of metaphor therefore rests on the ontological independence of the metaphorical image (the literal, unintended "sentence meaning" of the utterance) and as such testifies to the ability of verbal meaning to transcend the conditions of ordinary discourse—one person intending something to another within a shared context. The metaphorical image, unlike the metaphorical utterance, fulfills the condition of a literate text in positing a sentence meaning that exists apart from the immediate constraints of context and intention so crucial in determining utterance meaning.

The cryptic message in the bottle, on the other hand, belongs to the world of speech precisely because its sentence meaning is deficient and, because it depends heavily on deictic reference, can be meaningful only to someone who is physically present at the moment of utterance. The sentence meaning here has essentially no independence, no life, no room for play; instead it refers us directly to the situation that we share with the speaker and to the information that is embedded in the situation as part of our larger social relationship with the speaker. Its sentence meaning, then, exists only as a part of the utterance meaning, which in turn depends upon the larger social context. Since the sentence meaning exists essentially as a part of a larger social relation between the two communicants, we can label its function as *synecdochic*; in synecdochic communication, the literal meaning being communicated (what Searle calls "sentence meaning") exists only as part of a preexisting social context. The sentence meaning of the message in the bottle is largely synecdochic, since it has little value of its own, creating utterance meanings by referring to conditions that are largely shared or anticipated by the conversants as a result of social relations that exist apart from the message itself.

The status of Wilde's sentence is quite different, for here the ability of the sentence meaning to exist independent of context and intention gives it its standing as an aphorism. As with metaphor, the sentence meaning of Wilde's aphorism gains its power by first existing as an intervening or possible meaning between our first contact with the utterance itself and our understanding of its specific meaning within any one context. Our consideration of either the image of someone trying to climb a greasy pole or the sense in which price and value are different is not restricted by the demands of the moment. The playful element of metaphor and literacy comes from the fact that we are free to grapple with sentence meaning, as we are free in play generally, apart from the restrictions of ordinary social relations.

As the most perceptive analyses of metaphor have always recognized, the playful element ensures that the metaphor itself has a richer life than any literal paraphrase, since the intervening meaning is truly unconstrained—neither the writer nor the context ultimately controls the range of sensations, denotations, and connotations we draw from this meaning. The intervening sentence meaning, here the connotative richness of the image of climbing a greasy pole, with its appropriate blend of exertion, grime, and peril, thus establishes a life of its own, even as it partakes of a particular situation. Because its meaning transcends any one referent or the needs of any one situation, we can refer to the *symbolic* relation between possible meaning and actual context that lies at the heart of literate communication. Like the figurative image of a metaphor, literate verbal meanings generally attain the status of a symbol. In so doing they contain a certain element of the "magic" that Burke finds characteristic of a "perfect or total act," namely, "a modicum of novelty" that is not explainable in terms of scene, agent, agency, or purpose. Such an act, Burke concludes, must be in some respect "a motive for itself" (1969:66). The essential element in literacy, therefore, is not transcribed or coded language but symbolic verbal meaning, that is, language used to create meanings that are valuable to others independent of either immediate context or immediate intention.

The opposite of literacy, accordingly, is not simply oral language but synecdochic verbal communication, or context-bound language used to convey a specific intention in a specific context. *The opposite of literacy, in other words, is ordinary conversation.* As described by the philosopher H. P. Grice, ordinary conversation can be defined in terms of the implicit obligations to which we necessarily commit ourselves when speaking with others. Since conversation is a fundamentally social endeavor, Grice believes that the possibility for its success rests on our mutual support of four governing conditions: *quantity*, which obligates us to give enough but only enough information to sustain the dialogue; *quality*, which compels us to be truthful; *relation*, which obliges us to be relevant; and *manner*, which requires us to be perspicuous. All conditions work together to ensure that the conversants subordinate their own individual needs, which may conflict with the demands of the conversation itself. As Grice writes, the participants should "have some common immediate aim," and "each party should, for the time being, identify himself with the transitory conversational interests of the other" (1975:48). Accordingly, the verbal meanings transacted by the participants should always serve the needs of the conversation, and the meaning of the conversation can only be the mutual understanding reached in the pursuit of that common aim. Yet this common aim, in defining the purpose of the conversation, guarantees its synecdochic nature—whatever is said derives its value from serving the larger mutual needs of the two parties in their effort to reach accord.

Certainly conversations do fail, and probably because one or both parties fail to ensure the synecdochic character of their utterance, but such a point does not alter the fact that the goal of conversation is to create an understanding between two or more people actually brought together in a specific context. In its most basic form, conversation serves to draw two strangers together. It may ensure their lack of mutual hostility by having them agree on so common a topic as the weather—an ideal topic not only because it is not controversial and can thus almost guarantee agreement but also because it is necessarily part of the very context that any two people share and thus can be discussed largely by means of deictic reference. Even in the most exalted form of dialogue, however, the sort of Platonic dialogue between two souls envisioned by critic Denis Donoghue, what is said still serves some more compelling, mutual purpose: "What makes a conversation memorable is the desire of each person to share experience with the other, giving and receiving. All that can be shared, strictly speaking, is the desire: it is impossible to reach the experience. But the desire is enough to cause the reverberation to take place which we value in conversation" (1981:43).

Working against this need to subordinate ourselves to establish contact with another is our delight in creating verbal meanings that resist the pressures to reach social accord. Such meanings, embedded in such symbolic uses of language as metaphors and aphorisms, transcend the immediate needs of the conversants, even if they further a conversation. In this respect they perform in a general way some of the function that Freud assigned specifically to jokes—namely, "to promote the thought by augmenting it and by guarding it against criticism" (1960:133), that is, by elaborating and making acceptable what otherwise would likely not be uttered. This power of literate texts to resist the demands of conversation can best be seen in wordplays like puns and riddles that constitute the minimal form of literate activity. In violating all four conditions of the cooperative principle—quantity, quality, relation, and manner—such wordplays act as a form of what Burke calls "pure persuasion" (1973:269). As such, they involve the "saying of something, not for an extraverbal advantage to be got by the saying, but because of a satisfaction intrinsic to the saying." An act of pure persuasion, Burke adds, "summons because it likes the feel of a summons." Wordplays in general convey more meaning than is called for; while one meaning may serve the immediate needs of the conversants, the second meaning is necessarily excessive, thickening the verbal texture of the conversation and delaying the fulfillment of any immediate purpose. They thus illustrate Burke's point that "pure persuasion" can be maintained only by "interference" (p. 274).

The literate character of wordplays can be seen in a trick question that occasionally appears on intelligence tests: "I have two coins that equal fifty-five cents. One of them is not a nickel? What are they?" The key to responding correctly to this question has to do far less with innate intelligence than with one's general understanding of the literate use of language, that is, with uses that do not involve Grice's cooperative principle. Conversation as defined by Grice assumes the good faith of the communicants—the goal, after all, is shared understanding. Assuming the good faith of the questioner, we interpret this question as telling us that the "one" coin that is not a nickel is either coin. The trick, and hence the secret to responding correctly, is to adopt an entirely different psychological stance. No longer can we assume the good faith of the other party; we are not engaged in conversation. Instead, the question is a minitext intent upon objictifying a meaning, not in sharing it. Specifically, we err here in assuming the questioner's intention to communicate. Indeed, it is a riddle precisely because its synecdochic meaning, governed by our expectations, conflicts with its symbolic meaning.

As in interpreting metaphor, we backtrack, seeing the question as a symbolic utterance aimed at breaking through the confines of accustomed practice only when we realize that our initial understanding leads to an anomaly. At this point the answer suddenly appears: only *one* of the two coins is not a nickel—we can assume nothing about the *other* coin. The symbolic meaning here, like a metaphor's sentence meaning, has lain in silence, waiting to be discovered. We have been tricked, but we are also delighted to find once again that the range of verbal meaning is not limited by our current knowledge and our current expectations. There are ways of being in the world other than those to which we are accustomed and that we readily anticipate. Riddles, puns, and wordplays in general fulfill the function of all literate discourse in revealing to us that these new ways are readily objectified in language. Literacy can be defined, quite simply, as the act of reading and writing such language.

III. The pun, the metaphor, and the aphorism are all minimal units of literacy because their goal is less to facilitate mutual agreement between parties than to attain the status of a text by giving verbal meaning an identity free from the control of the cooperative principle. The crucial distinction at work here is not between any formal properties of written and spoken language but between writing and speaking as historical phenomena. Just as not all oral language is governed by the cooperative principle, not all written language strives to become a text. The formal properties of speaking and writing nevertheless do tend to reinforce differing functions; when we speak we are usually with others and thus ordinarily governed by the need to reach a mutual understanding, and when we write we are usually alone and thus ordinarily concerned with establishing meaning independent of immediate intent or context.

What Ricoeur says about spoken discourse is thus phenomenologically true, not logically necessary: "In spoken discourse the ultimate criterion for the referential scope of what we say is the possibility of showing the thing referred to as a member of the situation common to both speaker and hearer" (1976:34). We can always identify the actual conditions of a conversation—the "here," "now," "I," or "you" of any conversation. There is always additional shared information—the common background or purpose that occasions the conversation—which, while usually not physically present, also supplies a world of deictic reference, a world of shared meanings that can be indicated or minimally suggested ("You know what I mean!") without needing to be created in language itself. Yet even what is created anew in conversation must enter this deictic world, must ultimately be shared by the conversants, if it is to exist as anything other than as spoken words that are never heard. This "grounding of reference in the dialogical situation," writes Ricoeur, "is shattered by writing."

Beyond deictic reference, however, speech has even stronger ties to context. When we converse with others, not only do we convey meaning but we also act in the course of speaking to assert, question, commit ourselves to promises—we perform a whole array of activities called speech acts. Our words themselves have a meaning independent of these actions that is comparable to Searle's sentence meaning and to what Searle and other speech act theorists refer to as locutionary force. Likewise, our utterance can exert upon our listeners many different effects, called the perlocutionary force of utterances. Yet in speaking we also do things that are independent of either the content of our actions or their effect upon others. These activities—the assertions, questions, and promises already mentioned—constitute the illocutionary force of language.

Speaking is not only a means of conveying information but also itself a form of action, a way of asserting, questioning, promising, cursing, and performing a host of other actions. As the linguist Edward Sapir noted more than a half century ago, we often use language, not to create new meanings or even to indicate shared experiences, but as a form of action within a larger pattern of social interaction: "It is important to realize that language may not only refer to experience . . . but that it also substitutes for it in the sense that in those sequences of interpersonal behavior which form the greater part of our daily lives speech and action supplement each other and do each other's work in a web of unbroken pattern" (1956:9).

With regard to our speech acts, it is important to realize that the richer and denser the locutionary force—that is, to the extent that the locutionary meaning takes on the independent quality of a text—then the less useful such a meaning often becomes in serving the immediate illocutionary purpose of the speaker. When the attention shifts to what we are saying or, to be more accurate, to what is being said, then the illocutionary force of the utterance must be held in abeyance. We do perform many tasks in speaking, but to the extent that any one task is paramount, more pressing in importance than the locutionary force of the message itself, then the message necessarily acquires synecdochic value as part of a larger effort to achieve some particular aim in a particular situation.

Disraeli's reference to a greasy pole, Wilde's comment on price and value, a tester's puzzle about a nickel—all these utterances do neces-

sarily have illocutionary force, but their literate quality (the characteristic that each one shares with texts generally) comes in the delay that
each offers to the onward rush of conversation. In each case the
contemplation of the locutionary meaning at least momentarily draws
us away from the social demands that Grice sees governing conversation. "What writing actually does fix," notes Ricoeur, "is not the event
of speaking but the 'said' of speaking ... What we write, what we
inscribe is the noema of the act of speaking, the meaning of the speech
event, not the event as event" (1976:27).

Writing so defined involves a relationship between the individual
and the text that is fundamentally at odds with speech-based universal
models of discourse like the one shown below. In it literate texts are
presented as a special kind of message that passes between two persons, just as a writer is considered a special form of addresser and the
reader a special form of addressee. At the basis of such a model is the
insight that in some ways literate communication does follow a rhetorical paradigm—texts do serve as messages uniting two parties in
an act of communication. The problem here is that such explanations
often succeed at the expense of denying to literacy its defining quality—
namely our struggle, generally absent in conversation, as we are forced
to change in the process of creating or comprehending texts.

Addresser — — — — — — Message — — — — — — Addressee
(Writer)   (Text)   (Reader)

Generating the text is always problematic for the writer, just as
comprehending it is for readers. Rhetorical models based in speech
therefore generally lack enough of a psychological and historical dimension to account for this struggle. The text is not just a message
connecting two people; it is also a mediating agency for a single individual as he or she changes in response to the task of creating or
comprehending symbolic verbal meaning, as indicated by the diagram
below. Although certain aspects of literacy can be represented by
speech-based models of communication, other models of human interaction that draw on the notion of personal development or are based
upon notions of symbolic activity may more accurately reflect its distinctive qualities.

Reader/writer — — — — — The encounter — — — — — Reader/writer
prior to the   with the text   after the
encounter   encounter
with the text   with the text

Ricoeur, for one, draws upon the key term of German philosopher
Hans-Georg Gadamer, the "fusion of horizons" (*Horizonverschmelzung*),

to describe the ways in which the writer and the reader meet in the text—not as if they were exchanging messages but as if the most distant vision of each party overlapped: "The world horizon of the reader is fused with the world horizon of the writer. And the ideality of the text is the mediating link in this process of horizon fusing" (1976:93). In such a metaphor, the writer and the reader are necessarily separated. Neither party can have direct contact with the other, and thus there is no immediate context to complete or to dominate the speaker's intention. In a world where reader and writer are forever hidden from one another, we are obliged to assume less about intention and instead to weigh the message itself more seriously. For example, we are under greater obligation in such a situation to interpret the expression "I could care less" as a qualified assertion of concern or, if we do not, at least to view the implicit synecdochic ties to this writer, which allow us to ignore the sentence meaning and interpret the utterance meaning directly, as a diminution of the distance between writer and reader that is necessary for literacy. Such ties, we feel, violate the spirit of literacy by substituting a synecdochic relation for a symbolic one, allowing us to know what someone meant regardless of what he or she said and thus weakening, however slightly, the general power of language to objectify symbolic meaning.

The possibility of literacy resides in the very power of texts to distance us from writers and specifically from intentions that can be communicated apart from the text. To use another metaphor for literate exchange, also drawn from Gadamer, as readers we meet the author in the text in much the same way that we encounter someone through the agency of a game. All our personal knowledge of our opponent is helpful in planning our strategy only to the extent that this knowledge is actually relevant to our opponent's play of the game. Even if this opponent is a close friend, we have no assurances that what we know about her strengths and weaknesses off the court will be expressed in her play. The game, for example, could be a diffident person's single outlet for aggression. Similarly, we can learn something of an opponent's personality through observing her play even if she is a total stranger. And here the similarity with literacy is most compelling, for in such a situation we do learn the personality of the other as competitor, just as all the reader knows is the personality of the other as writer. "In this sense," Ricoeur notes, "Heidegger rightly says . . . that what we understand first in a discourse is not another person, but a 'pro-ject,' that is, the outline of a new way of being in the world" (1976:37).

Thus just as we never truly play a game against a friend, only an opponent, so we can never know a writer directly but only as a reader. The agency of the game or the text transforms a synecdochic relationship into a symbolic one. In a synecdochic relationship, rhetorical motives dominate; participants fully engage in the activity not for its own sake (as they should in any game) but only in order to reach some more encompassing social accord (as we might lose a game on purpose to curry a favor). Similarly, we can write with the clear intention of using the text as a means of reaching a desired social goal; our activity is purely symbolic, however, only when our social relations with others are disregarded in the performance of the new activity that establishes its own relationships between parties. The game exists only as a symbolic activity in the participants' mutual denial of what is given (namely, their existing social relation apart from the game) for the new way of being in the world that is constituted by the rules of the game itself; that is, we are only playing a game when we play to win, regardless of how we might act toward our opponent ordinarily. "Play," explains Huizinga in *Homo Ludens*, "is not 'ordinary' or 'real' life. It is rather a stepping out of 'real' life into a temporary sphere of activity with a disposition all of its own" (1955:8).

If in playing the game we abandon all concern with the effect of our actions on our social relationship with the other, we must also abandon concern with intention as well. We can certainly intend many things when we play, and in turn we can infer the intentions of our opponents, just as we can intend things when we write and infer the intentions of authors as we read, but all such intentions and inferences in themselves are relatively inconsequential; that is, they matter only to the extent that they are expressed through the public acts that constitute the formal play of the game. This public nature of games, moreover, allows us, within the rules of the game, to forgo our normal concern for the other's well-being, for example, to avoid being solicitous, thus providing the source of psychological freedom that lies at the heart of such rule-constituted activities as games and literacy; both take place apart from our normal social obligations.

To the extent that literate communication is rule governed, it may appear to be just another form of what Searle and others define as speech acts. Rules for speech acts, however, are closely aligned with psychological intention. Our performance of the speech act of promising, for example, closely parallels our psychological intention to commit ourselves to some obligation. In more purely symbolic activity, where meaning has little to do with what someone intends by an act, the question of sincerity or felicity plays a far less important role. While we may be curious to learn what someone actually had in mind when performing a certain rule-governed action (for example, what a dancer had in mind during a performance or what a quarterback had in mind when throwing an interception in a football game), such information has little to do with the meaning or even with the play of the game itself, provided that the quarterback is indeed playing the game (and is hence necessarily engaged in a symbolic activity) and is not *playing at the game*, as he would be if he were playing poorly on purpose in order to fix the outcome of the contest.

Clearly our understanding of someone's intention plays a major role in determining the meaning of nonsymbolic or direct social actions, for example, in determining whether a remark was meant as a compliment or as an insult or in determining whether the act of taking someone's life was premeditated or accidental. Yet with largely symbolic activity, we forgo such direct concern with intention in order to concentrate on the play of the game. Here, as "readers," we can establish an understanding of our opponents' game, of their strengths and weaknesses as players, and even of their likely moves that far exceeds even their knowledge about themselves. Our insight is not into their private intentions or into their hoped-for achievements but into their motives, that is, our constructions of their intentions based upon a set of public actions. Because motives, unlike intentions, are finally public, interpreters of human behavior generally, whether opponents in a match or readers of a text, can have a greater understanding of the mediating agency of the game or text itself, including the ability to anticipate future direction, than the person originating the action.

The speech-based models of communication fail to account for literacy, since the text, like the game, is not a vehicle for something already in the writer, either as message or as intention. Like athletes, writers express, not their intentions, but themselves in their work. There is no athlete except through the game, just as there is no writer except through the text. While the writer can communicate to others through the text, just as an athlete can perform for an audience, in neither case does this rhetorical consideration outweigh the fact that the area of contact between performer and spectator is the symbolic activity itself. When we read or write literate texts, we are embodying meaning in language much as an athlete creates the meaning that is the play of the game.

In even so minimal an element of literacy as the visual pun "Sᴑrolls," for example, we encounter only the crafter of the pun, only the playful intelligence capable of objectifying the meaning "Dead Sea Scrolls" in such a form that its comprehension will be for us an act of discovery of hidden meaning. We can discover no more information about the author through the text, and any information we may acquire from other sources may well have nothing to do with our reading of this text. This visual pun therefore reflects the fundamental distinction Ricoeur makes between "ordinary language," whose purpose is "to build a bridge between two spheres of experience," and "scientific language," whose purpose is "to insure the identity of meaning from beginning to end of the argument" (1978:129). The goal of this pun is not communication but an identity of meaning independent of circumstance.

As a minimal work of literacy, the visual pun acts as a text, not by connecting the reader and the writer, but by acting for each as a self-sufficient, intervening agent between the self that is and the self that will be. This intervention of the text within the psychological life of the individual results in the reflective nature of literacy—as writers, we can know ourselves only through the mediating agency of the text; that is, only as readers can we know ourselves as writers. Since reading necessarily involves the interpretation of public meaning, it is possible to assert that all writers, even great authors, have only greater intimacy with their work and with the intentions expressed therein. There is no privileged knowledge of texts themselves.

Speech-based models of discourse, as they attempt to encompass the unique features of literacy within a general model of communication, have no mechanism for accounting for the objectifying, transcending power of texts. While any study of communication will be part of an identifiable rhetorical context, its value as a work of analysis is apt to depend in part on the extent to which its insights into communication are themselves relatively free of the very rhetorical conditions it seeks to analyze. Thus the very possibility of rhetoric depends upon the objectifying power of texts and hence upon the historical emergence of literacy. As Walter Ong notes, rhetoric is "the 'art' developed by a literate culture to formalize the oral communication skills which had helped determine the structures of thought and society before literacy" (1971:49).

It is hardly surprising that one of the earliest uses of literacy was to study orality, whereas the formal study of literacy is a comparatively modern phenomenon—indeed, one might argue, a largely contemporary one. More surprising, however, is the persistence of the effort to explain literacy largely as a special class of speech. Speech-based models, as compared with those based upon other meaningful human activities, such as play, art, cognition, and symbolic activity generally, explain only issues that are peripheral to literacy.

IV.   Literacy is the ability to read and write as defined in terms of the ability to understand and create symbolic verbal meanings stored in some permanent form. Such a definition of literacy rejects an unproblematic distinction based solely on the means of transmission and instead substitutes a model that is problematic in two distinct ways. First, it defines literacy as the antithesis of the unproblematic model, calling it the ability to read what one would never hear and to write what one would never say. To be literate, accordingly, is to be able to "speak" and "listen" to a whole new class of people—those whom we can know only through texts.

This new model of literacy is also problematic in a second, more profound sense. In shifting the definitional burden from coding to meaning, it ensures that the specific judgments regarding literacy will always remain open to argument. There is no longer a basis for making a fairly straightforward distinction between everything that is spoken and everything that is written, between the world of the ear and the world of the eye, and, in so doing, keeping debate over definition at the level of a border dispute. Rather, the crucial distinction is now between language used as part of the daily interpersonal relations and language used to establish an identity of meaning independent of such social interaction. The former is largely the domain of speech and the latter the domain of writing, although not exclusively.

The differences that matter most are therefore between literacy and conversation, not between oral and written language, and these differences exist in relative degrees rather than in absolute categories. Reading is something like listening, writing is something like speaking, and texts are something like utterances. Accordingly, it is possible to trace the history of the reading of any one text, to reveal its synecdochic ties with certain groups, and in so doing to emphasize the connection between literate and ordinary communication. Readers of texts are members of interpretive communities, but the differences between reading and speech communities are at least as important as their similarities.

It is not possible, for example, to equate the life of the text to its history with any set of readers, as an utterance might be described in terms of those who heard it, since the very act of historical reconstruction gives the text a new and expanded life, and a new and expanded audience in the present. Texts in the form that reaches us as readers are, in Fredric Jameson's words, "always-already read" (1981:9); our response to anything we read, like our involvement in any conversation, is necessarily constrained to some extent. The relative degree of constraint, however, remains crucial. Reading remains phenomenologically distinct from listening, among other things, because of the increased likelihood that the characteristically greater privacy brings a certain freedom from the immediate social pressures to conform. Reading matter always has the potential to conflict with people's inherited sense of things. Compared with listening, reading is more likely to encourage people to adopt a critical stance toward their own experiences. The literate experience of reading always contains elements of rejection and renewal.

The act of reading thus loses its synecdochic connection with a specific context to the extent that we as readers attempt to understand what a text has meant and continues to mean to others. Texts have synecdochic rather than symbolic ties to readers, in other words, only as history itself, in Jameson's words, is "the ultimate ground as well as the *untranscendable* limit of our understanding in general and our textual interpretations in particular" (1981:100)—that is, only to the extent that the totality of our understanding, however encompassing it may be, is itself part of the very world we are studying. All acts of reading and writing do take place within limiting historical conditions, but our involvement with literacy, if nothing else, is part of our effort to expand our sense of the present.

In literate acts of reading and writing texts, we are trying to break away from synecdochic ties to any one historical moment, in part by reinterpreting the very world we inhabit. While all human understanding, no matter how far reaching, is ultimately a part of the very historical context in which it is created, the very acts of reading and writing play an important role in expanding our sense of who we are and where we live. The idea that literacy can be defined in terms of the mastery of a finite set of texts, as suggested by E. D. Hirsch in his program for "cultural literacy," therefore involves a fundamental contradiction in terms, an attempt to restrict the real power of reading by pedagogic force. In Ricoeur's words, our reading of texts gives us a "world and not just a situation," a *Welt* and not an *Umwelt:* "It is this enlarging of our horizon of existence that permits us to speak of the references opened up by the text or of the world opened up by the referential claims of most texts" (1976:36, 37).

Literacy, then, is nothing less than our ability to deal with discourse that projects a world—to deal, in other words, with the world of texts. In this sense, literacy remains the ability to read and write, but the meaning of these key terms changes. Whereas transcription may remain a facet of literacy, at least for the foreseeable future, the defining criterion of literacy is one's ability to create and comprehend the symbolic verbal meanings of texts. When we write to someone because, like students under the watchful eye of a teacher, we cannot talk, the text we produce is likely to be no more than a substitute for speech. Yet the note as writing carries with it the potential for becoming a literate text to the extent that it is capable of unfolding a new meaning or a new way of being in the world to a reader.

Although some letters seek to reaffirm a mutual understanding against the threat of separation, others, like Kafka's letter to his father, strive to objectify meaning independent of a preexisting relation, indeed to shatter and reformulate a relation upon new grounds as set forth in the text. To read a personal communication as a literate text is to read it as a commentary on one's relationship—a commentary that even when it acts to reaffirm the status quo does so on its own terms. To read such a commentary, even from one's closest friend, we must first comprehend the symbolic meaning that is embodied in the text—where we meet, not necessarily the friend we have known, but the author of the text, the friend we are coming to know.

The fundamental obstacle to reading, therefore, is not decoding but attaining the ability to overcome synecdochic expectations. In the words of Wolfgang Iser, the reader needs the power "to open himself up to the workings of the text and so leave behind his own preconceptions" (1974:291). Above all else, in other words, readers require the psychological motivation to separate themselves from what they share with others. Here Iser quotes George Bernard Shaw on the inherent tension between synecdochic and symbolic understanding: "You have learnt something. That feels at first as if you have lost something." Decoding skills may be a prerequisite for reading, but at some point readers must be willing to alienate themselves from synecdochic meanings.

Thus to the extent that people feel content with their current situation or see little purpose in acquiring a new understanding, then reading as the comprehension of texts will have little value, and we can expect such people to have little success in school, even after years of reading instruction. Such students will have little regard for the necessarily strange and distant world to which we lay claim in the very act of comprehending. Ricoeur calls reading the "*pharmakon,* the 'remedy,' by which the meaning of the text is 'rescued' from the estrangement of distanciation and put in a new proximity" (1976:43). Students struggling to learn how to read, however, often lack not just the "remedy" but also, as an initial malady, the sense of estrangement from the present that prompts us to seek other answers. As reading theorist Frank Smith notes, students learning how to read confront only two basic obstacles: their recognition, first, that print is meaningful, and second, that it is different from speech.

Even these two points, however, really rest upon a single basic assumption, since print is ultimately meaningful because it differs from speech. In the final analysis, the single greatest obstacle to reading is the students' failing to acquire a compelling interest in verbal meanings that differ categorically from those normally communicated in their social relations with others. As an aid in overcoming this obstacle, reading to very young children, even to infants, attains great importance, for such reading attaches feelings of emotional security to the necessarily threatening acts of comprehending that which is strange and different.

If reading involves overcoming the estrangement of verbal meaning, then writing must be the mirror image—that is, the distancing of oneself from the pressures to reach social accord by embodying new ways of being in a text. Writing is less the means of transferring encoded verbal meanings to readers than the effort to create that meaning and objectify it in the form of the text. In some vital sense, we do not usually create the meanings that we communicate in conversation; they are often ours by virtue of our living with others. In our conversations we are usually content to facilitate our being in this world by negotiating accommodations with others, often by means of deictic reference to a common condition.

In *After Babel,* George Steiner argues that the great multiplicity of languages in the world, nearly all of which have existed only in spoken form, is the direct result of this important but often ignored function of language—to preserve the distinctive collective experience, the synecdochic meanings, of the group: "Each tongue hoards the resources of consciousness, the world-pictures of the clan. Using a simile still deeply entrenched in the language-awareness of Chinese, a language builds a wall around the 'middle kingdom' of the group's identity. It is secret towards the outsider and inventive of its own world" (1975:232). Or as Sapir noted more directly, "The mere fact of a common speech serves as a peculiarly potent symbol of the social solidarity of those who speak the language" (1956:16).

Given the synecdochic pull of speech, it is hardly surprising that, however few people ever learn to read texts, far fewer can ever write them. Writing is a fundamental act of social betrayal; Steiner insightfully locates the source of literacy—what he calls "vital acts of speech"—in private experience that, from the point of view of the group, represents lying, falsehood, and saying "the thing which is not"—saying, in other words, what is not ordinarily called for by a given context. Like the image of the metaphor, the text thus violates what is normally called for in any given situation; indeed, the text is a metaphorical image raised to the level of discourse—it is the literal lie that expresses a greater, figurative truth, even when the text purports to be factual, historical, or scientific.

All texts, fictional and nonfictional, are in this sense opposed to conversation—they all seek to reformulate an existing relationship with the world. Steiner talks about language that preserves collective identity and that which transforms it—conversation serves the first purpose and literacy the second. This latter use of language, Steiner writes, "*is the main instrument of man's refusal to accept the world as it is.* Without that refusal, without the unceasing generation by the mind of 'counter-worlds'—a generation that cannot be divorced from the grammar of counter-factual and optative forms—we would turn forever on the treadmill of the present. . . . Ours is the ability, the need, to gainsay or 'unsay' the world, to image and speak it otherwise" (1975:217–18). In making permanent such "deceits," writing fulfills in literacy the creative impulse latent in all language use for what Burke calls the "peculiar possibility of the Negative" (1966:420). "Only writing," Ricoeur adds, "in freeing itself, not only from its author and from its ordinary situation, but from the narrowness of the dialogical audience reveals this destination of discourse as projecting a world" (1976:37). The writing of a story, in this sense, completes the motive that led to its telling.

2

Modern Literacy
Education

I.   Where do our modern notions of literacy come from? What is the source of our current expectations concerning reading and writing achievement? Upon what ground is the widespread sense of deterioration based? We do know that, by the middle of the nineteenth century, the United States had one of the lowest overall rates of illiteracy in the world. The early immigrants to this country were often better educated than the countrymen they left behind, and once settled in the New World, their Protestant zeal for reading the Bible and their democratic and commercial aspirations ensured a steady growth in education from the mid-seventeenth century on. By 1850, while Europe had an illiteracy rate of 50 percent, or 60 percent if Russia is included (Cipolla 1980:71), the first census in the United States to count illiterates on an individual basis showed an illiteracy rate of only 22 percent. Since this figure included the total slave population of the South as illiterate (Fogler and Nam 1967), the overall nonslave illiteracy rate was less than 11 percent and approached zero for certain large subgroups, such as male New Englanders.

These figures, however, indicate only the percentages of people who could not read at all. They tell us only that, on the basis of certain historical records, most male New Englanders during a certain period could sign their names, had attended school for a minimum number of years, or met some equally narrow criterion. In other words, they refer to "literacy" only in its unproblematic sense. Although it is difficult to say with certainty just how well or how poorly people could actually read and write, it is possible to be more precise about the amount of schooling—quite little by today's standard—that formed the basis of "literacy" in mid-nineteenth-century America. One source of such information, a survey of 1,200 New Englanders in the Northern army between 1862 and 1864 (Soltow and Stevens 1981), reveals that, while only 3 percent had no schooling whatsoever, only 7 percent had attended high school for even two years, and only 1 percent had ever attended college. Thus the near-universal literacy of New Englanders meant that some 90 percent of the males had attended elementary school for an average of 5.2 years—admittedly more than their counterparts in any other region and perhaps enough to make adults capable of signing their names and reading certain texts but probably not enough to guarantee that all students could comprehend anything unfamiliar, much less create texts on their own.

A half century later, the national average for years of schooling had increased significantly to 8.1 years, leading us to expect a comparable decrease in the national illiteracy rate, and indeed the rate did fall drastically from the 22 percent indicated by the 1850 census to the 7.7 percent reported by the census of 1910. Such statistics, however, say little about the "literate" activities that the other 92.3 percent could actually perform. One indication of their level of literacy is given in the extensive data collected by psychologist Robert Yerkes in his controversial program for testing the intelligence of all U.S. Army recruits in 1918—a group that, because of its relative youth, should have been even better educated than the general population. This census, however, determined "literacy" merely by asking whether or not each person over the age of ten was able to read and write. Less than 8 percent of the population, in other words, considered themselves illiterate, and only passive acceptance of the unproblematic model of literacy would allow anyone to assert that over 92 percent of the population was therefore fully "literate" in some more compelling sense.

Yerkes, for all the shortcomings of his work in intelligence testing, could not afford to be so uncritical in his approach to literacy. Although he had no interest in studying literacy per se, he nevertheless wanted to give all recruits a written test and thus had to confront the practical problem of determining just which of the recruits were in fact *literate* enough to take the written, or alpha, form of the test and which had to take the nonverbal, or beta, form, administered with oral instructions. Clearly, the ability to take the alpha form constituted proof of literacy more rigorous than that accepted in the 1910 census, and Yerkes accordingly did see the need to incorporate some crucial, although minimal, elements of the problematic model into his definitions. The test for illiteracy now included not just lack of attendance for more than one or two years at an American public school but also the inability either to read or write personal letters or to read a newspaper written in English (1921:745).

In applying these criteria, Yerkes discovered, as we might well expect, that many people deemed *not illiterate* by the standards of the U.S. census were also "not literate" by his own, more demanding criteria. Indeed, of the million and a half men that Yerkes tested, one-fourth were unable to meet some version of his test for literacy and were thus given the beta, or oral, form of the test. An additional 6 percent were given the beta form after failing the alpha test, and Yerkes reports his frustrations at his inability, for logistical reasons, to retest all the recruits who scored so low on the alpha test as to raise questions about the validity of the original determination of literacy. Thus, while the 1910 census reported a rate of illiteracy of less than 8 percent, Yerkes found a rate of over 30 percent for an all-male, younger, and presumably better educated subgroup. New England recruits, although outperforming groups from other areas, still fell within the same general pattern, with 403 men of 1,537 from Massachusetts in one sampling, or 26 percent, having to take the beta test.

Yerkes was interested in establishing a scientific basis for his views on the hereditary nature of intelligence, and the wealth of data produced by his massive testing program has subsequently attracted attention in large part because of its racist overtones and its role in encouraging changes in the immigration laws. Although Yerkes's research was hopelessly faulty as a measure of native intelligence, it nonetheless has inestimable value as an indicator of the great gap between common perceptions and new expectations. With Yerkes's work, for the first time in U.S. history, literacy was defined for a significant cross section of the population at least partially in terms of the ability to negotiate symbolic verbal meaning. The results, reflecting the limited reading and writing skills of a nominally educated populace, proved to be a revelation. As one surprised educator wrote, of those soldiers unable to take the alpha test, "an overwhelming majority . . . had entered school, attended the primary grades where reading is taught, and had been taught to read. Yet, when as adults they were examined, they were unable to read readily such simple material as that of a daily newspaper" (Burgess 1921:12).

Given the exceedingly low expectations that are built into the unproblematic model of literacy, especially as compared with the specific skills needed to take a written test, it seems obvious today that no one should have been surprised at Yerkes's results (and it is even more amazing that anyone would have been foolish enough to use the results of such a test to make "scientific" claims about innate intelligence). But to many professional educators of Yerkes's generation, the connection between decoding skills and the ability to comprehend seemed so inevitable that they were indeed unprepared for the early results. The problem was comparable, not to listening and still not comprehending (as might be the case when we strain to follow the reasoning of a speaker and still do not understand), but to being able to hear what someone in authority says and still not minding him or her, an action readily explainable only in terms of some social or genetic defect such as obstinacy or stupidity.

As one researcher wrote in 1925, surveying test results of typical ninth graders, "The average reading comprehension as compared with the total comprehension possibilities of the selections used is so mediocre that it is very hazardous to proceed on the assumption that students in the ninth grade can read well enough to understand and appreciate literature merely by reading" (quoted in Applebee 1974:90–91). Again, it should hardly have been surprising that fourteen-year-olds who have mastered decoding skills might still have difficulty understanding the plays of Shakespeare, the prose of Macaulay, the poetry of Browning, and the novels of James Fenimore Cooper—the works used in this study and the common fare of the English curriculum of the day. Such results were surprising, however, because the appearance of written tests requiring people to read something unfamiliar signaled a radical departure in educational expectations, a departure of which few people at the time were aware. As historian Daniel Calhoun commented, the standards embodied in these written tests, "whatever their limitations, showed that men were creating a mental role to which they might hold the performance of ordinary Americans. This standard had not existed a hundred years earlier" (1973:78).

In *The Education of a People*, Calhoun surveys the emergence of new cognitive demands in American life in such diverse activities as child rearing, preaching, and shipbuilding in the second half of the nineteenth century. The change in educational practice, especially as regards reading and writing instruction from the 1860s till the 1910s, has had perhaps an even greater impact upon contemporary life. At the start of this period, the teaching of reading and writing was based almost entirely upon the unproblematic model of literacy: in the limited schooling available to most students, reading was taught as decoding and writing as encoding.

Reforms inherent in the problematic model existed, then, largely in the theoretical discussions of certain educational writers. By the end of the period, the situation had largely been transformed, with certain reforms already well entrenched in the language curriculum and others the subject of heated controversy. By the third decade of the twentieth century, students in the United States were regularly struggling to understand unfamiliar texts and, at least in progressive school districts, were laboring even harder to create their own. During the half century between 1860 and 1910, wide-scale literacy education as we know it today first came into existence. It is therefore no coincidence that, while the terms "illiterate," "literate," and "illiteracy" are hundreds of years old, the first recorded use of "literacy" appeared in an American education journal of 1883, in the very middle of this crucial fifty-year period.

II.   Reading instruction in the United States throughout most of the nineteenth century was based upon the oldest and seemingly the most basic method—teaching students the names and then the sounds of the alphabet as they learned to recite some quasi-official texts. This method is exemplified in the famous *New England Primer*, first issued in the late seventeenth century and reprinted continually throughout the eighteenth century. It required students to learn the alphabet in conjunction with religious and didactic verses such as the child's prayer "Now I Lay Me Down to Sleep." Such an alphabet-based, mnemonic approach has at its center a profound dichotomy between the mastery of decoding skills and the mastery of content. On the one hand, the mastery of decoding skills entailed an atomistic, incremental approach to learning, in which students started with the sounds of the alphabet and then tackled the sounds of ever longer units. According to one pedagogue in 1821, "The pupil commencing with the alphabet ought to be taught his letters at sight, until he knows them perfectly, and then the sounds that arise by combining letters; proceeding gradually from one stage to another, until he can read words of one or two syllables, without spelling in the columns, as they are generally placed in our spelling books" (quoted in Soltow and Stevens 1981:97).

On the other hand, the content of children's reading matter had no incremental or developmental component whatsoever, no direct connection with either their mastery of decoding skills or their interests in the world around them. The *New England Primer* asked the child only to recite a rhymed, rhythmic statement of some widely held belief, as in the following selection:

> I in the Burying Place may see
>   Graves shorter there than I;
> From Death's Arrest no Age is free,
>   Young children too may die;
>
> My God, may such an awful Sight,
>   Awakening be to me!
> Oh, that by early Grave I might
>   For Death prepared be.

Here are adult sentiments in adult language for a child to read or, more accurately, to recite in public. There is no attempt to have children read words that decoding skills at their level might allow them to pronounce; learning the sounds of letters and reciting texts are fundamentally distinct processes. While we may wonder whether children understood the text, the curriculum gives no indication that comprehension was a vital or even necessary part of reading instruction. Rather, as in the teaching of a catechism generally, the teacher was to explain the meaning of the text to the student in conjunction with the student's actual recitation but separately nevertheless. Indeed, the recitation was often an exercise more in mnemonics than in decoding, and students educated by such instruction were apt to feel that the goal of reading instruction was to enable someone to "read" from memory or, as an English hand-loom weaver bragged, to teach people to "say th' catechis fro' end to end, and ne'er look at the book" (Altick 1957:168–69).

While the proportion of religious and moralistic passages did change throughout the nineteenth century, with reading matter becoming in the process more secular and more nationalistic (Applebee 1974:1–5), the emphasis remained on oral recitation of material that students by themselves would ordinarily have trouble decoding, much less fully understanding. Such is certainly the case with all of McGuffey's Readers, including the last edition of the sixth and most advanced volume in 1879, which, in addition to the texts (more appropriately labeled "selections for oral recitation"), had supplementary material on the principles of articulation, inflection, accent and emphasis, voice, gesture, and the reading of verse—all areas of elocution that have generally disappeared entirely from the curriculum in the last hundred years.

In terms of the development of the theory and practice of reading instruction, the McGuffey Readers, although barely a hundred years old, more closely follow the pattern of language instruction found in traditional societies that have little in common with late nineteenth-century America. The method used has more in common, for example, with reading instruction in mid-nineteenth-century Egypt, where, as one British traveler noted, the student memorized first the alphabet, then the ninety-nine epithets of God, and finally the opening chapter of the Koran (Goody 1968b:262). The student then proceeded to learn the other chapters, beginning with the last and working backward. In such traditional societies, according to Jack Goody, the pedagogical principle is decidedly different: "The elementary class is taught not to read but to recite, simply using the letters as mnemonics for what comes next" (p. 222). Goody was reporting on his observations in northern Ghana, but R. B. Ekvall, another commentator, saw an even stronger connection between reading instruction in twentieth-century Tibet and ritualized recitation:

The illiterate cannot use a book in his praying, except to hold it in his hands and raise it to his forehead to "rub off the blessing" as he intones the syllable *OHm*. The literate, on the other hand, derives greater benefit by being able to scan the lines and turn the pages, for such an activity is also part of verbalization. He visualizes the meanings in their written form and holds them in his mind. Such comprehension constitutes an added degree of observance. [1964:125]

Despite the added status that accrues to "readers" of the text, their understanding of the world and that of their illiterate companion remain fundamentally the same. Neither is engaged in the task of comprehending symbolic meaning.

Nor for the most part were students in English schools so engaged, at least as indicated by a University of Edinburgh professor who wrote in an American educational journal in 1830:

> English reading, according to the prevailing notion, consists of nothing more than the power of giving utterance to certain sounds, on the perception of certain figures, and the measure of progress and excellence, is the facility and continuous fluency with which those sounds succeed each other from the mouth of the learner. If the child gathers any knowledge from the book before him, beyond that of color, form, and position of letters, it is to his own sagacity he is indebted for it, and not to his teacher. [Quoted in Matthews 1966:55]

Implicit in this summary is the criticism that reading pedagogy is wrong to ignore so completely the question of the student's understanding of content.

While concern with the special problems that students must overcome in order to understand texts does play a major role in the efforts of educational reformers, it is important to realize just how unusual such concern really was within traditional pedagogic practice for much of the century. When in 1839, for example, the British educational reformer Sir James Kay-Shuttleworth prepared a list of six minimum requirements to be met by candidates for a teacher training program, the reading requirement stated only that the successful candidate must be able to "read with fluency, ease, and expression" (Kay- Shuttleworth 1839:81). Meanwhile, the one explicit test of comprehension of a verbal text asked the candidate to recite the catechism from memory and then to explain its meaning, with the clear understanding that successful candidates were to exhibit the orthodoxy and not the originality of their understanding.

The assumption here, as in reading instruction generally in traditional societies, is that the most important verbal understandings of individuals are readily available to all, regardless of anyone's mastery of coding skills. People thus had to understand the same things whether or not they were capable of decoding writing. Kenneth Lockridge makes essentially this point in surveying the impact of literacy in colonial America: "There is no evidence that literacy ever entailed new attitudes among men, even in the decades when male literacy was rapidly spreading toward universality, and there is positive evidence that the world view of literate New Englanders remained as traditional as that of their illiterate neighbors" (1974:4).

Certainly symbolic verbal knowledge was contained in books and periodicals to which individuals with coding skills alone had access. Any sense that one learned to decode in order to acquire this knowledge, however, seems to have been lacking. Accordingly, there was little awareness that reading itself was the act of comprehending meanings that are not ordinarily shared with one's neighbors in the normal course of direct social interaction. As Daniel Calhoun notes in *The Intelligence of a People*, the second half of the nineteenth century in American life witnessed both a rapid expansion in knowledge, as people became increasingly skilled and specialized in diverse areas, and a marked increase in anxiety and insecurity about just how little any one person actually knew and could do to control the future. The more we learned and the more intelligent we appeared to become, the less we really seemed to know. Calhoun contends that, prior to the industrial expansion following the Civil War, American life made few purely cognitive demands. Few activities required the mastery of intellectual skills apart from those ordinarily learned in the general process of socialization. Even schooling, which today we tend to consider as involving largely the mastery of cognitive skills, was for the most part devoid of measurable symbolic content in the first half of the nineteenth century, at least according to Calhoun:

So long as pupils were familiar with the school process and with the language in which they were taught—so long, that is, as they did not suffer from massive "cultural deprivation"—teachers could hardly conceive of any learning blocks internal to the specific learners they were encountering. They could not, for a simple reason: between the irregular pace at which schools were conducted, and the modest subject-matter that parents expected or permitted teachers to offer, there was no extensive schoolroom transaction that served as a test of learning ability. [1973:75]

"Reading," regarded as the decoding of familiar material, certainly did not constitute such a transaction.

Indeed, as long as educators felt that synecdochic meanings were far more important than symbolic ones, then reading pedagogy would rightly be more concerned with both decoding and elocution than with comprehension. Beginning in the 1820s, however, certain Americans, influenced by European educational reformers, began to promote a revolutionary new approach to reading instruction—one that attempted to integrate children's mastery of decoding and their comprehension of content by having them learn from the outset to sound out meaningful verbal units, that is, whole words. As one such early reformer, John Miller Keagy (1792–1837), wrote, the child "should by no means be taught his letters, or spelling at first, but *whole words* should be presented to him, to be pronounced at sight" (quoted in Mathews 1966:64–66). No longer would decoding be taught in terms of mastering discrete, atomistic sounds, while students recited complex, linguistically unrelated texts.

The reform advocated by Keagy represented not just a change in pedagogy but a fundamental shift in the goal of reading, with attention focusing for the first time on the task of awakening in the child the meanings embodied in texts. Reading, quite simply, was to change from a branch of mnemonics to a branch of cognition, and if children were to become adept readers according to this new standard, their curiosity must first be aroused so that they would want to learn more about the world around them. Reading instruction, therefore, must begin not with decoding and memorization of texts but with the arousal of students' interest in symbolic knowledge. With this in mind, Keagy advocates that the classroom for the first two years of schooling be a miniature museum filled with objects that would stimulate interest. "In such an establishment," Keagy writes, "teaching to think and to understand, would be found to be a much more delightful and easy occupation, than the usual one of teaching not to think."

A decade later, Keagy's call for the reform of reading pedagogy was picked up by the great Massachusetts educator Horace Mann, but even Mann's status was insufficient to withstand the protest from a teaching profession still married to the belief that reading was a branch of elocution designed to help people in their public avowal of widely shared truth. By the end of the nineteenth century, however, the disparity between a pedagogy designed to produce mechanically efficient decoders and the wider needs of society for students with the ability to comprehend meanings that were not regularly negotiated in the home or any other familiar social setting had grown much greater, thereby affording many opportunities for a new generation of educational reformers.

Indeed, the immediate, practical need for students' greater proficiency in symbolic exchange had become so obvious that reformers had merely to indicate the existing conditions to make their case—as did physician-turned-educational-journalist Joseph Mayer Rice in publishing a series of articles in the *Forum* from October 1892 to June 1893. The principal target of these articles, reporting on conditions in thirty-six cities and published as *The Public-School System of the United States*, is what Rice calls "purely mechanical drudgery"—the fetish for drill and superficial mechanical correctness that Daniel Calhoun sees as one of the most common and most ineffectual cultural responses to the widespread perception that much higher levels of cognitive performance were being required from Americans as a people. Everyone seemed to sense that schools had to increase the ability of all students to think symbolically, but the only pedagogy many educators could imagine was an intensified version of that used in an earlier and quite different era. The frantic quality of teaching at the time is evident in the words of one Chicago teacher who, standing before her class, inveighed, "Don't stop to think—tell me what you know!" Rice, in particular, found "concert recitations" an especially odious form of pedagogy, "preeminently fitted to deaden the soul and convert human beings into automatons" (1893:88–89).

Some of Rice's few words of praise were for the work of Colonel Francis Parker of the Cook County Normal School. Colonel Parker, later to be the director of the School of Education at the University of Chicago and there an early supporter of John Dewey, had been a leader in educational reform at least as far back as his eight-year tenure with the public schools in Quincy, Massachusetts (1875–1883). At the center of his reforming efforts lay his clear commitment to a new, problematic model of reading:

The mere pronunciation of words, however correctly and readily done, is not reading as here defined. The teacher who concentrates effort upon the vocal utterance only, or upon the vocal utterance first and the thought afterwards, is leading her pupils astray. Her teaching is formal, and not real. The all-important habit for the child to form is that of never supposing that a sentence has been read before the thought is clear in his mind.

The main point, therefore, to which the attention of the teacher should be directed at every step, from first to last, in the teaching of reading, is this: *Are the pupils led to get the thought?* [Quoted in Mathews 1966:106]

Such an unambiguous assertion of the problematic model of reading, however, should not be confused with its total or even wide acceptance. Colonel Parker had numerous detractors both in Quincy and later in Chicago, and the debate as to how to teach reading continues essentially unabated even today—as in some sense it must, given our ambivalence about the dual role of "reading" in both reinforcing prevailing attitudes and promulgating understandings that will strike many of one's acquaintances as anomalous. The idea that the goal of reading is the comprehension of meanings not immediately shared by those with whom we live, and especially by the parents of children, inevitably breeds its own opposition. Nevertheless, certain aspects of the problematic model of reading did become so widely incorporated into the theory and practice of reading instruction in the late nineteenth and early twentieth centuries that it is still possible to speak of a true paradigm shift in our understanding of literacy. Two changes, in particular, have proved to be so pervasive that, looking back today, it is somewhat difficult to imagine that they were not always part of reading instruction.

The first of these two changes was the transformation of reading instruction from primarily an oral activity to primarily a silent one. As Louis Kelly notes in his survey of twenty-five hundred years of language arts instruction, "Until the twentieth century there is hardly a mention of silent reading" (1969:152). Reading instruction was construed as an oral activity for thousands of years, yet the new emphasis on silent reading during the last two decades of the nineteenth century seemed both so sudden and so natural that even as astute a student of reading pedagogy as Edmund Burke Huey had trouble accounting for the persistence of the earlier tradition.

As Huey noted in his pioneering work of 1908, *The Psychology and Pedagogy of Reading*, "Reading as a school exercise has almost always been thought of as reading aloud, in spite of the obvious fact that reading in actual life is to be mainly silent reading. The consequent attention to reading as an exercise in speaking . . . has been heavily at the expense of reading as the art of thought-getting and thought-manipulating" ([1908] 1968:359). Why, Huey wanted to know, does pedagogy in the first decade of the twentieth century continue to stress oral recitation, when the goal of reading is clearly understanding? The simple answer, of course, is that what struck Huey as so obvious—namely, that reading is essentially a means of "thought-getting"—had only just become plain to forward-looking educators. His proximity to this change robbed him of a proper historical sense of what was happening.

Writing fifteen years later, another educator, Henry Grove Wheat, had an easier time providing a historical perspective on this issue. Wheat was specifically able to identify four "social needs of former days" that required what he called "the teaching of expressive oral reading": (1) reading material was scarce; (2) only a few could read; (3) communication was slow; and (4) speech was the chief means of communication (N. Smith 1939:159). Likewise, he continues, "the social needs of the present require the teaching of rapid silent reading": (1) reading material is abundant; (2) almost everyone can read; (3) communication is rapid; and (4) writing is the chief means of communication.

Here Wheat has described a number of the symptoms of the shift from an unproblematic to a problematic model of reading, but he does not mention the one overriding issue. Whereas in the past the most important verbal meanings were shared by the community, and one read aloud as a means of attesting to one's acceptance of these meanings, by the early twentieth century, much significant verbal knowledge, upon which, for example, economic well-being depended, was specialized, technical, innovative, and private; that is, it was not shared by the group as a whole. One now read, not in public to affirm one's beliefs, but in private and in silence to gain command of that new knowledge.

Once students had begun reading silently in order to gain access to symbolic verbal knowledge, a new pedagogic question arose—how should successful reading be judged? Reading competence had traditionally been measured entirely in oral performance, but Colonel Parker's contention that reading is essentially "the art of thought-getting and thought-manipulating" makes it conceivable for students to be decoding properly and still not reading because they are somehow not following the sense of the text (conversely, it also becomes conceivable that someone can study how to *read* a foreign language and not be able to decode it properly). The real test of reading, therefore, is understanding. Here we have the second great transformation in reading pedagogy—the widespread use of written tests, often in standardized form, of reading comprehension.

That the goal of reading is acquiring the knowledge necessary to answer questions about the material read may at first seem so self-evident as to make it hard to believe Louis Kelly's assertion that we have no records of any systematic effort to assess reading comprehension before the early decades of this century (1969:35–36). At the higher levels of education, people were tested, albeit often in oral form, about knowledge that may have been most directly gained through reading, but before the twentieth century, only a tiny minority of people fell into this category. What was not measured in colleges and universities or anyplace else was the *generalized* ability of students to extract meaning from a text dealing with material about which they had little prior knowledge. In large measure the reason was that no one believed that people really learned anything when they learned how to read. Reading was for the most part a means of ensuring a more personal access to religious truth. Today one may too easily forget that before the rise of the modern industrial state in the second half of the nineteenth century education for most people in the world was narrow, limited, and sporadic.

As already noted, New England recruits in the Northern army averaged only 5.2 years of schooling by midcentury, and the figures were considerably lower in other regions. Nearly 20 percent of the recruits from Kentucky, Tennessee, the coastal slave states, and Canada had never attended school, while nearly half had done so for only three years; thus only about a third of the recruits from these areas had more than three years of education. Furthermore, in other countries as well only very limited formal schooling was available to the population as a whole. Nowhere in the world in the mid-nineteenth century, not even in Sweden, a country with a record of almost universal literacy going back to the seventeenth century, was there a widespread system of formal education before 1850. In the absence of schools, the Swedes, for example, learned to read at home by reciting aloud—where they were hardly likely to have their reading comprehension evaluated systematically in writing.

Until the great expansion of state-funded education in the second half of the nineteenth century, the attainment of even marginal decoding skills was a significant accomplishment. In *Literacy and Development in the West*, the Italian historian Carlo Cipolla cites one official report after another showing that only small percentages of working-class children in fact attained such skills. A report from Naples in 1829, for example, notes that, of 2,000 girls attending school, only one-fifth learned to read properly (that is, to decode). Another report from Turin, the province in Italy with the lowest rate of illiteracy, reports that in the late 1870s there was still "no rational system of education, which does not extend beyond a little reading and learning to write badly. Results are poor. After a few years many pupils are no longer able to read a document or even to write their name intelligibly" (1980:35). In England, the Factory Act of 1802 required the millowners to provide schools for child workers, yet the results were only marginally better than those from Italy; an 1837 report, for example, records that, of 2,000 students attending nineteen different factory schools, only 933 could decode the Testament and a third of these only "with difficulty" (p. 67n).

The standard test of reading competence for most of the century, as I have mentioned, merely asked students to read aloud familiar material. Such a reading test was in fact incorporated into the Revised Code, which provided for minimum-competency testing in state-inspected schools in England between 1862 and 1897 and for which Matthew Arnold served as a state inspector for twenty-four years. Specifically, students in grades one to five were required only to read aloud a brief passage selected from their textbook, while sixth graders were required to read a paragraph from a newspaper or narrative (although in 1882, partly at Arnold's urging, sixth graders were also asked to read a literary selection). Nowhere, however, was there any provision for the state inspector to test, either in writing or orally, the students' comprehension and Arnold constantly complained in his annual school reports that the students actually understood very little. Before the Revised Code had run its life in England, American educators were showing considerable interest in tests designed to compare oral and silent reading comprehension, but not until 1915 was the first written standardized test of reading comprehension, the Courtis Test, available to teachers. As I noted at the beginning of this chapter, the startlingly low levels of understanding revealed by this and other such tests first drew wide public attention to the basic difference between decoding and reading comprehension.

The introduction of systematic written tests of reading comprehension may at first seem to have to do less with the emergence of the problematic model of literacy than with the triumph of scientific methodology; after all, such tests emphasize standardized results. The initial concern with the problematic nature of comprehension, however, has a quite different origin, beginning not with educational psychologists late in the nineteenth century but with Romantic thinkers early in the century and specifically with their realization that hermeneutics—formerly restricted to problems relating to biblical translation and exegesis—was in fact a general problem in the reading and understanding of any text. Hermeneutic philosophers first grasped that all written texts, not just the Bible or ancient manuscripts, create problems of interpretation, since they embody historical experiences that are fixed in time. At the foundation of our modern idea of reading, therefore, lies Schleiermacher's recognition that "there is hermeneutics when there is misunderstanding" (quoted in Ricoeur 1981b:46)—the recognition, in other words, that texts are by nature uprooted from the very situations that facilitate their ready comprehension, and thus understanding them entails essentially the same skills as the effort to understand forms of cultural experience different from our own.

While we can readily decode texts without understanding them, we can only *read* a text if we can grasp its unfamiliar content. Written tests of reading comprehension thus measure, admittedly often quite crudely, the extent to which students can substitute the understanding embodied in a text for their own. The standardized testing of reading comprehension may have originated in nineteenth-century positivism, but the tendency to consider comprehension problematic does not. While certain aspects of the problematic model of reading do remain controversial, there is nonetheless widespread acceptance of two revolutionary nineteenth-century pedagogic views—the belief that reading is principally a silent activity and that success should be measured in terms of comprehension of content. Reading as we know it today in large measure reflects a limited triumph of the problematic model of literacy in the last hundred years.

III. Although reading instruction in the United States remained limited for much of the nineteenth century, it was still far more ambitious than writing instruction. The terms "reading" and "writing" today may seem complementary, but they have very different pedagogic histories. As I have already noted, societies based on certain traditional patterns of organization will often regard the main purpose of literacy as aiding people in the public recitation of belief and, as such, will have little use for writing. In nineteenth-century Egypt, for example, only the few students destined for careers as clerks were even taught to write, and such instruction was concerned only with the proper formation of letters and other skills needed in record keeping.

Similarly, S. J. Tambiah in his account of education in Thailand reports that writing was a branch of calligraphy. In one village Tambiah studied, the monks, who were largely responsible for the transmission of literacy, apparently never wrote a new text or even a new commentary on an old one. "The texts," Tambiah notes, "became 'fixed' and comprise a sacrosanct body of knowledge, transmitted 'unchanged' through time" (1968:94).

It is indeed impossible to understand the magnitude of the cultural transformation that lay behind the emergence of the writing curriculum in twentieth-century America without considering the situation elsewhere for much of the nineteenth century. Even in Sweden, the most literate country of the time, less than 20 percent of the population had any demonstrable encoding skills at the beginning of the nineteenth century, and less than half the population was able to encode by midcentury (Johansson 1981).

The actual writing pedagogy practiced within formal schools for most of the nineteenth century is evident from the two "writing" requirements established by Sir James Kay-Shuttleworth for prospective teachers. They were asked to "write in a neat hand with correct spelling and punctuation, a simple prose narrative read slowly to them" and to "point out the parts of speech in a simple sentence" (Kay-Shuttleworth 1839:81). We today are likely to associate such exercises in dictation and grammar more with introductory classes in a foreign language than with writing classes in English. Formal writing instruction in one's native language consisted mainly of work in penmanship and spelling and included, at somewhat more advanced levels, the study of prescriptive grammar (as represented in Murray's textbook, which was based upon complex and, for English speakers, often unnatural Latin models). At least at the elementary level of schooling, beyond which few students ever progressed, there was no instruction in English composition for much of the nineteenth century.

As with reading, however, the eventual basis for the reform of writing pedagogy was provided in part by changes in cultural orientation that were associated with the Romantic movement of the late eighteenth and early nineteenth centuries but with one important difference. The study of hermeneutics expanded throughout the nineteenth and twentieth centuries from the study of biblical exegesis to the study of the problems involved in understanding texts and eventually, with Heidegger and Gadamer in this century, encompassed the general problem of time-bound human understanding. The Romantic interest in the process of generating texts, however, tended from the outset to restrict itself to identifying the properties of poetry, a special class of writing.

For the Romantic tradition, poetry differed from other forms of language, since it alone attained the ideal character of a text, what Kant referred to as "purposeness without purpose" *(Zweckmassigkeit ohne Zweck)* or what, in terms of the problematic model of literacy, might be called "meaningfulness without intention."

In the lyric poem, as nowhere else, a verbal text shows that it can possess value apart from the immediate concerns of either speakers or listeners. In Wordsworth's famous expression, poetry has its origin in "emotion recollected in tranquility," that is, apart from active involvement with the world. As Meyer Abrams observed in *The Mirror and the Lamp,* his study of Romantic poetics, "the paramount cause of poetry is not, as in Aristotle, a formal cause, determined primarily by the human actions and qualities imitated; nor, as in neo-classical criticism, a final cause, the effect intended upon the audience; but instead an efficient cause—the impulse within the poet of feelings and desires seeking expression, or the compulsion of the 'creative' imagination which, like God the creator, has its internal source of motion" (1953:22). The poem, as such, is a Romantic metaphor for the text.

As certain forms of verbal expression like lyric poetry gained in stature in the early nineteenth century (in large measure because they epitomized symbolic rather than synecdochic communication), other forms, specifically those involving the direct engagement of other people, were losing stature. After all, if the most valued form of verbal expression is that which originally emanates from the "spontaneous overflow of powerful feelings," later "recollected in tranquility," then verbal expression that is precalculated and uttered in the midst of some pressing concern would seem to be necessarily less valuable. Thus the corollary to the rise of the poem as the ideal text is the diminution in the standing of rhetoric—the form of language study that since classical times had stood at the center of language study in higher education. While such rhetorical training formed the basis of higher education the study of written language was generally confined to the earliest years of schooling in what were called "grammar schools."

In conceptualizing the poem as a self-sufficient text, Romantic poetics drew attention away from the dynamics of rhetorical exchange and to the question of how certain kinds of verbal meanings are able to exist, in Ong's words, "free of the dialogic stuggle with an audience" (1977:222–23). The Romantic poem, Ong continues, is "no longer a riposte but a simple product, an 'object' rather than an exchange," and here he quotes John Stuart Mill's aphorism—"Eloquence is heard; poetry is overheard." While the speaker is free to impress others, the writer must be concerned entirely with establishing the self-sufficiency of the text, that is, with ensuring that the text will retain its symbolic communicative function independent of any one context. Any other effort, either to affect an audience directly or to use the text for some ulterior purpose, diminishes its ontological status. Mill, writing here as a spokesman for this new Romantic sensibility, is emphatic on this point: When an "act of utterance is not itself the end, but a means to an end—viz. by the feelings he himself expresses, to work upon the feelings, or upon the belief, or the will, of another—when the expression of his emotion . . . is tinged also by that purpose, by that desire of making an impression upon another mind, then it ceases to be poetry, and becomes eloquence" (quoted in Abrams 1953:25). Poetry is thus elevated to the status of text seeking to establish an ideal meaning: eloquence, on the other hand, is criticized for using artistry in the service of some more immediate, practical goal.

The relative reversal in the standings of rhetoric and poetry eventually led to specific pedagogic changes in the nineteenth-century language arts curriculum, changes that were comparable to, although perhaps somewhat less dramatic than, those in reading pedagogy. Rhetoric, for example, in large measure transformed itself from the study of speechmaking to the study of college theme writing. The lectures of Edward T. Channing, professor of rhetoric at Harvard University between 1819 and 1851, reveal the basis for this transformation, which in the second half of the century provided the theoretical foundation for an entirely new college course of study, English composition. Like Mill, Channing is very distrustful of language that aims for effect. An audience, he says, is "not assembled to be subjects upon which [the orator] may try the powers of his eloquence, but to see what eloquence can do for the question. The subject is more thought of than the author, and what he says must come from the subject rather than from his art. . . . The splendor that surrounds him must be the natural light of truth, not the false brilliancy that startles and blinds" (1856:17).

The principal goal of communication is not convincing others but establishing what is true. Conviction that results in the deep agreement of speaker and listener apart from the truth of the text is a form of blindness. The image of the most exalted form of communication, therefore, is no longer that of the orator moving a gathering but that of a solitary individual engaged with a text: "What nobler or more original study can we have than that of a man in his writings—a contemplative being—and often more fully disclosed to us there, than he could be in a life of action; and brought near to us for our minute inspection and full sympathy" (pp. 196–97). As Wallace Douglas comments in his study of these lectures, at stake here is a "quite radically new view of rhetoric": "Its paradigmatic image is that of a man thinking, of a man whom 'increased knowledge' has made 'more contemplative'" (1976:116). With Channing, the orator becomes the man of letters.

While Channing was laying the groundwork for the college study of composition, the actual curriculum at Harvard still reflected its classical heritage. Throughout the 1850s the term "English" at Harvard continued to refer to the study of elocution and rhetoric and consisted of a four-year course of study that led students from lessons in orthoepy and expression as freshmen to the study of forensics as seniors (Grandgent 1930:81). Only in 1873 was elocution made an elective and the first writing requirement installed for incoming freshmen. In wording, this requirement, which asked each candidate to "write a short English composition, correct in spelling, punctuation, grammar, and expression," did not, however, represent a radical departure from the past or the total triumph of the problematic model of literacy, since the purpose of the exercise was less to create a text than to create a correct model of a text. The requirement's main purpose was to determine whether students could produce replicas of proper texts. Accordingly, in addition to exhibiting all the proper forms of expression, the successful Harvard applicant had to demonstrate in the essay sufficient knowledge about the standard authors and their works, as drawn from an official Harvard-approved reading list.

To some extent, then, the Harvard writing requirement can be viewed as both an extension and a rejection of rhetoric, which throughout the century had been increasingly a study of the formal properties of literature and thus naturally expanded to include the study of the formal properties of compositions about literature. In order to complete the writing assignment successfully, the student needed to demonstrate no innovative or personal understanding of the material under consideration—orthodoxy in interpretation and correctness in expression, each in accordance with a prescriptive canon, were the rule. The basically conservative thrust of the tradition is apparent from the guidelines of the Harvard-oriented New England Association of College and Preparatory Schools concerning the classroom work necessary to ensure that students have command of "correct and clear English, spoken or written." Specifically, these guidelines called for "training in grammar and the simple principles of rhetoric, and the writing of frequent compositions." The candidate was:

to spell, capitalize, and punctuate correctly. He must show a practical knowledge of the essentials of English grammar, including ordinary grammatical terminology, inflections, syntax, the use of phrases and clauses; a thorough training in the construction of the sentence; and familiarity with the simple principles of paragraph division and structure. [Quoted in Stewart 1976:50]

Nowhere is there any guarantee, or even any concern, that the essay show an advance in understanding on the part of the reader (as one might fairly expect only from a professional literary critic) or that of the writer (as one might regularly expect from a bright student responding to probing questions). While the resulting composition is thus clearly not speech, it is also not likely to be a literate text; it is instead a curious anomaly—the classroom composition, or work written to demonstrate a certain level of proficiency in the mechanics of English composition.

This rhetorical tradition emphasizing correct responses to questions on a set literary canon had far less influence outside New England. In other areas of education, in public education especially, there was more pressure to abolish the study of rhetoric entirely and to adopt pedagogic practices based upon the problematic model of literacy. One English educator who did call for the total abolition of formal rhetoric was Samuel Thurber. He argued that we want students to demonstrate not knowledge of rhetoric itself but the effect of having such knowledge, that is, good writing (Carpenter, Baker, and Scott 1913:221). There does seem to be less reason to study the formal properties of good writing if the goal of instruction is the production of a text that is finally to be judged, not on its mechanical features, but on the ability of writers to embed a new understanding of themselves or of their world in a text.

Not before the early twentieth century, however, did educators regularly and openly offer sweeping affirmations of the problematic model of writing, as in this statement from English educator Edward Holmes writing just three years after the publication of the formal guidelines of the New England Association in 1908: "I mean by composition the sincere expression in language of the child's genuine thoughts and feelings. The effort to express himself tends, in proportion as it is sincere and strong, to give breadth, depth and complexity to the child's thoughts and feelings and through the development of these to weave his experiences into the tissue of his life" (quoted in Mathieson 1975:61). To become a writer, the child must master any number of formal skills, but Holmes points composition pedagogy in an entirely new direction by subordinating all such skills to the ultimate purpose of objectifying experience.

By 1911, the sentiments that Holmes expressed were beginning to have a direct impact on language education. In that same year, some fifty English educators gathered in Chicago to form a professional organization, the National Council of Teachers of English, in large measure as a protest against the unwillingness of college educators to free themselves from a conservative rhetorical tradition and to champion a new model of literacy (Hook 1979). The founders of the NCTE saw themselves as representing the interests of public secondary education and state universities against the efforts of newly formed college associations, which, dominated by eastern private schools, were attempting to establish a uniform college-preparatory curriculum. The writing component of this curriculum was to consist of student essays that would reflect the reading of a select list of literary classics. The founders of the NCTE were in part bolstered by the democratic spirit of the Midwest, the adolescent psychology of G. Stanley Hall, and the educational philosophy of John Dewey, but in their fight against a rhetorical, model-based writing curriculum, they were also bolstered by a new view of reading and writing.

James F. Hosic, one of the founders and most prominent early leaders of the NCTE, headed a joint committee, cosponsored by the National Education Association, that in 1917 solidly endorsed this new model of writing. The report, entitled *The Reorganization of English in Secondary Schools*, attacked the traditional college preparatory curriculum for requiring students to write on a narrow range of literary topics that happened to interest teachers and for not treating the students' own experiences as suitable subjects for compositions. Students, Hosic argued, should be encouraged to write about their interests—field trips, hobbies, sports, current events—all activities that "touch the life of the student in some way":

The development of the expressional powers of the individual pupil should be the aim of the teacher rather than the teaching of specific form and rules. Each year of a pupil's life brings a broader outlook through added experience and more mature thought. Each year, consequently, there is need for an increased mastery of technique and of more mature forms of expression. Only from a realization on the part of the teacher of this growth of personality can an adequate course in composition be organized. [1917:54]

Hosic's claim is radical in the history of literacy finally because an educator was advocating that the material to be transcribed for permanence was not sacred texts or important legal documents but ordinary individual experience.

This viewpoint gave student writers and their teachers an entirely new pedagogic problem. Whereas a lack of proper encoding skills or grammatical knowledge or understanding of some prescribed text had previously been the main obstacles to success in the writing curriculum, now students also had to develop the power of reflecting on their own experiences. As with the shift in reading signaled by the reforms of Keagy, writing was to become a branch of cognition. Before it could do so, the writer necessarily had to evince a heightened curiosity about his or her place in the world:

To sum up: In the composition course, content should appeal to the pupil as first in importance; organization, second; details of punctuation, spelling, sentence structure, choice of words (matters of careful scrutiny). third. . . . In general, the classroom activities in composition should spring from the life of the pupil and should develop in him the power to express his individual experiences. [Pp. 57, 59]

Just as the adoption of the problematic model of reading led to educators' realization that students could decode properly without actually reading, so the adoption of the problematic model of writing, as here advocated by Hosic, brought the comparable realization that one can encode properly without writing—certainly a common complaint of college composition teachers today. Hosic was asking students, as good writing teachers still do, to perform no mean feat: they must objectify their own experiences in written language so that the resulting compositions interest readers with whom they have no direct social connection. Student writers today, in a word, are expected to become the creators, not just the reproducers, of texts.

IV.   The pedagogic changes in the teaching of reading and writing in American schools during the nineteenth and early twentieth centuries were accompanied by the dramatic rise of English as a new academic discipline. Access to postelementary education for most of the last century, as I have already noted, was severely limited in this country and throughout the world, but the minority who did attend high school and college as a rule did not "study English" nor did they spend much time reading and writing texts written in English. In higher education for most of the century, language education focused mainly on rhetoric, classical language and literature, and, somewhat later, philology. Pre-college education placed additional emphasis on grammar. At both levels, however, the principal concern reflected the dominance of faculty psychology—the belief that faculties of the mind such as reason and memory are best trained by the rigorous study of subjects that, no matter how irrelevant to current concerns, can be formulated into a complex series of formal rules. The study of Latin, for example, was considered valuable apart from its content because it developed mental skills that would contribute to learning generally.

As long as reading and writing were considered exercises in coding and not modes of understanding, there was little chance that a discipline concerned with "reading" and "writing" vernacular texts could ever gain much standing in the curriculum. Instead English had to enter the schools under the aegis of a traditional discipline with a well-established, complex method of investigation. Accordingly, at the college level, the study of English literature was initially a branch of philology, and the greatest emphasis was given to the most philologically interesting texts—namely, those written in Old and Middle English. At the precollege level, the study of literature became a branch of national history (often an exercise in the memorization of dates), while the study of language itself was a branch of traditional, Latin-based grammar. Of all the possible "legitimate" covers for English in the nineteenth century, rhetoric, with its concern for identifying and cataloging the characteristics of literary texts, was perhaps the most congenial.

Although the discipline of English did meet with less resistance in this country than in England, the history of its progress in the curriculum is almost entirely confined to the second half of the nineteenth century and even then was severely limited prior to the last two decades. One survey of thirty-four Massachusetts secondary schools reveals that only fourteen had any offerings in vernacular literature in 1867, with twelve more adding courses in the 1870s and the remaining eight schools adding them in the 1880s. It must be stressed, however, that such courses were often electives and dealt largely with the study and memorization of facts contained in literary histories. In an examination on Milton, for example, one school in 1866 asked students to sketch Milton's life to 1638, to outline "L'Allegro," giving examples of obsolete words, and to copy from memory passages from "Il Penseroso" and "Lycidas," indicating from the former "which words in the passage are from Anglo-Saxon, which from the Latin" (quoted in Applebee 1974:29). Schools, complained educational reformer and critic John Churton Collins in 1887, too often regarded literature "not as the expression of art and genius, but as mere material for the study of words, as mere pabulum for philology . . . Its masterpieces have been resolved into exercises in grammar, syntax, and etymology" (1887:644).

Collins may have been discouraged, but at least in the United States, the force of events was to be on his side. In the next three-quarters of a century, English became the cornerstone of American education. A number of factors were at work here but most fundamental was the pervasive recognition that texts, even if written in the students' own language, are not fully meaningful to students without serious study. Inasmuch as mastery of decoding skills does not ensure that adolescents can understand Shakespeare, appreciation of our literary heritage makes necessary systematic instruction in English and American literature. A national leader in the drive to place the new discipline of English at the center of a revised educational curriculum relevant to the needs of a newly urbanized, newly industrialized America was Harvard University President Charles W. Eliot. Eliot lamented the "woeful ignorance of their own language and literature which prevails among the picked youth of the country" (1909:99) and thus found little reason to justify the study of English as a form of mental discipline. After all, he reasoned, Americans need both a command of written English and a knowledge of their literary history. Yet, when he surveyed educational practices of the day, he was shocked to discover that as late as 1876 the nation lacked "any definite, well-organized system of secondary instruction in the mother-tongue" (quoted in Carpenter, Baker, and Scott 1913:46).

Eliot took two steps that proved instrumental in establishing English as a regular academic subject. First, he led Harvard in 1873 to establish the entrance requirement in English composition as I noted above, and second, he was chairman of the widely influential Committee of Ten, charged by the National Education Association in 1892 with charting a new course for secondary education. While the committee's report and Harvard's entrance requirements both had strong conservative aspects, the two moves taken together propelled English forward as the single most important high school subject and gave it status at least equal to that of traditional disciplines in colleges and universities. In particular, the Conference on English, which met at the request of the Committee of Ten and had as its chairman the progressive Samuel Thurber, played a crucial role in recommending that English be taught in all four years of high school (as it still is in most parts of the country) and in issuing a succinct, two-part definition:

The main objects of the teaching of English in schools seem to be two: (1) to enable the pupil to understand the expressed thoughts of others and to give expression to thoughts of his own; and (2) to cultivate a taste for reading, to give the pupil some acquaintance with good literature, and to furnish him with the means of extending that acquaintance. [*Report* 1894:86]

English was thus defined as the academic embodiment of a new model of literacy: reading is understanding the "thoughts of others," and writing is giving "expression to thoughts of [our] own." In the last decade of the nineteenth century, President Eliot called it a marvel that within one generation a subject that "never had any place at all down to 1873, when it first appeared in the Harvard requirements," (1909:376) could have become the keystone of our national education.

For most of the nineteenth century, traditional disciplines such as philology, rhetoric, grammar, and national history all vied to control and to profit from the ever-increasing interest in a new model of literacy. The rapid rise of English as an academic subject free from these older disciplines largely resulted from the wide perception that the reading and writing of vernacular texts were indeed difficult tasks for most students. In this sense the problematic model of literacy possibly explains the growth of English best. Yet the strength of this new model is itself the result of deeply rooted historical changes. It is more revealing, therefore, to trace the rise of both English and the problematic model of literacy to the passionate need of intellectuals generally over the last two centuries, namely the urge to seek truth in experiences that were unconstrained by the seemingly all-encompassing, hostile forces of ordinary social life. Forced to live in a world that appeared increasingly unstable and subject to rapid and uncontrollable change, intellectuals throughout the nineteenth century deliberately sought to cultivate a critical spirit that would allow them to read, to write to, and, in a sense, to live with as kindred spirits, the writers of the past and the readers of the future. Intellectuals in the nineteenth century generally sought, in other words, a greater sense of the unknown and of history generally.

One expression of intellectuals' pervasive dissatisfaction with the effects of rapid historical change is Matthew Arnold's definition of culture as a continuous struggle between symbolic understanding of the world, available only to those capable of knowing "the best which has been thought and said in the world," and synecdochic practice based upon "our stock notions and habits, which we follow staunchly but mechanically" ([1868] 1961:48). Culture, accordingly, is not a specific form but a general attitude, in Arnold's words, "not a having and a resting, but a growing and a becoming." The study of literature in this context is really a form of reading enlivened by a critical spirit; we read great writing so that we may turn, in Arnold's phrase, "a stream of fresh and free thought" (p. 6) upon the world around us. Intellectual life constantly challenges us to reevaluate our synecdochic ties. Culture is therefore tantamount to resistance to the status quo. As F. R. Leavis, Arnold's twentieth-century heir, wrote, in conjunction with Denys Thompson, "We cannot, as we might in a healthy state of culture, leave the citizen to be formed unconsciously by his environment; if anything like a worthy idea of satisfactory living is to be saved, he must, he must be trained to discriminate and to resist" (pp. 3–4).

English as the study of literature ultimately reflects this spirit of resistance to the constraints of synecdochic practice, an attitude fully manifested in the pioneering pedagogical and critical work of I. A. Richards. *Practical Criticism* (1929), for example, is primarily concerned with showing that nominally well educated readers asked to comment on unidentified poems of greatly varying quality were able to respond in only a very limited way. Richards's point is quite clear—our ordinary responses are apt to be woefully inadequate when we are dealing with the complex demands of great literature. In an earlier work, *Principles of Literary Criticism* (1925), he had explained that great literature most essentially provides us with complex new ways of looking at the world.

The cliché, or stock, response, accordingly, is the enemy of critical reading, just as the complex short poem in embodying meaning free of a speaker's intention and context is the ideal text and hence the opposite of normal, referential uses of language. Richards, like Arnold, emphasizes the distinction between ordinary and special communication that lies at the heart of literacy, the difference between what Richards calls "ordinary life," where "a thousand considerations prohibit for most of us any complete working-out of our response," and "imaginative experience," where "these obstacles are removed," allowing us to see previously "unapprehended and inexecutable connections" (1925:237). The poem at the center of this "imaginative experience" is for Richards a pseudostatement, less because it is untrue than because, like the image of a metaphor and symbols generally, it is "free from belief." To explicate a complex lyric poem, ideally at least, is to *read* the poem in a sense far removed indeed from the notion of "reading" as the public avowal of faith, which uses the book as a mnemonic aid.

To Richards and other critics of his generation, what mattered most about the act of reading was neither the truth of the referent nor the effect of the message on any audience but instead the complexity of the intervening text, its irony and ambiguity, its richness of imagery, its interplay of texture and structure—all of the elements that, taken together, establish the text as a symbolic object. Scholars who viewed poems as such ideal objects were rightly called New Critics, for they were in fact embodying the problematic model of literacy in what became the first aesthetic of academic English. Literary critics for a half century have debated many aspects of New Criticism, but only rarely have they sustained any attack upon its two fundamental ties with the problematic model of literacy—that texts act to separate readers and writers and that reading and writing are means of revising our cultural heritage.

In the first case, the problematic model of literacy supports the notion that readers should not consider an author's intention relevant in understanding a text as long as that intention is not embodied in the text itself and is known by the reader only as part of a direct or indirect social relationship with the author. Whereas the prominent Victorian critic Leslie Stephen once stated that the goal of literary study was "to make a personal friend of the author"—in Stephen's words, so that he might be able to "recognize him if I met him at a railway station" (1887:487)—our goal as "new" readers is just the opposite—we seek what is hidden and complex in texts, and indirectly in their authors, even if they are written by people we see and with whom we converse every day. In the second case, the problematic model supports the New Critical assumption that reading and writing are valuable to the extent that they embody new understandings or new readings of what others accept uncritically. Thus any act of reading, if it is to be at all valuable, must contain an element of "deconstruction," if only of existing interpretations. In this sense, we must recognize that certain key aspects of New Criticism and hence of the problematic model of literacy have been extended and more deeply embedded in academic English, rather than supplanted, by nearly all successive schools of literary criticism.

Since the triumph of the New Critics over half a century ago, the problematic model of reading has remained at the center of English studies. The status of writing within the profession, however, has never been as fully decided. One reason is that rhetoric, concerned as it is with correct models of discourse, never entirely lost its place within the college writing curriculum. In part because rhetoric as early as the eighteenth century had already begun establishing itself in colleges as the study of correct forms of expression, the creators of the undergraduate curriculum have been reluctant to embrace fully the message of the Hosic report, that students are to be creators of texts.

Another obstacle to the success of the problematic model of writing in academic English has been the tendency to see the poem as representing not the symbolic nature of any literate text, but a qualitatively distinct category of verbal meaning. The aesthetics of academic English, for a number of reasons, has generally been content to limit the application of the problematic models of reading and writing to a very narrow and clearly defined category of texts labeled "literature." One result of this schism in the ontological status of literary and nonliterary texts is that different models of literacy apply to students as readers and as writers. As composition theorist Janet Emig notes, for most English professors only literature really qualifies as a "text"; the next lesser category is school-based creative writing, the "pseudoliterature" of would-be writers, followed in turn by the "non-literature" of the professors' own writings about literature; and least of all, the "pseudo-nonliterature" of students writing about literature in English classes (1983:173–74). Thus the model that governs the reading of literary texts would not ordinarily be applied when papers written by students are read by their professors.

In voicing more than forty years ago a lament of the profession that is often heard today, William Lyon Phelps reflects the dilemma that grows out of restricting the problematic model of literacy to certain categories of texts—those written by "authors" and not by students or even by professors themselves: "That a man should graduate from Harvard with honours, spend two years in advanced study in the Graduate School, then a year of research in Europe—only to correct spelling, grammar, paragraphing, etc., seemed to me a hideous waste of time and energy" (1939:273). Although such graduate study no doubt included a heavy dose of philology. Phelps's own interest as a teacher made him a constant advocate of the problematic nature of literary texts, leading him as a young instructor to offer one of the first English courses in the contemporary novel.

As a "reader" of literature, Phelps was prepared to judge texts according to new standards; nevertheless, as a "reader" of student compositions, he was content to judge texts in terms of a rhetorical tradition largely concerned with the question of the student's correctness according to preestablished models. Student writers are not seen as extensions of the great writers whom professors devote their lives to studying—no one really expects students to create symbolic texts. Rather they are seen as extensions of the professors themselves, and their writing is to be judged, as is most often the professor's writing, in terms of its ability to demonstrate mastery of prescribed academic conventions.

The fact that such research papers are often about literary texts and the writers of such texts only adds to the ambivalent position of English professors—those readers most attuned to the richly symbolic nature of certain kinds of texts spend much of their professional lives writing their own essays and reading the essays of students according to rhetorical-based standards of appropriateness. These texts, by definition, can never be as valuable as the literary texts that are being studied. The professors must learn certain formal conventions to progress within the profession and in turn must teach students other conventions in what are often deemed "service courses," but whether they represent "service" to the university, to the students, to society generally, or to literacy itself is unclear.

The result, as might be expected, is a disturbing division of focus within the profession, with professors of Romantic literature, for example, teaching, not the model of writing developed in Coleridge's *Biographia Literaria*, but the five-paragraph essay, types of logical fallacies, and other remnants of a neoclassical rhetorical tradition. The same professor who praises the descriptive writing of Ruskin in literature class is apt to be a staunch defender of the traditional hierarchy of the college writing curriculum, which affords argument and research a higher standing than description and narration. It is one thing to recognize the necessity of adjusting a curriculum to the needs and abilities of students; it is another matter entirely not to be able to see a connection between students and "authors" as writers.

William Riley Parker (1967), in his historical sketch of the growth of college English departments, refers to the "tenuous" connection between the teaching of writing and the study of literature within universities as the unavoidable result of their differing origins. Composition, he feels, has much closer ties with rhetoric and speech than with literature, and to the extent that the older rhetorical tradition has continued unabated in English departments, he is correct. Parker fails, however, to recognize the close historical connection between the sudden widespread rise of interest in the study of vernacular literature in the second half of the nineteenth century and the emergence of a whole new model of writing that was fully endorsed by progressive forces within secondary school education only in the decades between the two world wars.

The rise of English within the secondary school and college curriculum is in large measure the result of society's general acceptance of the problematic model of literacy. The inability of college English departments to integrate the study of reading and writing is finally the result of a failure to perceive the common origin of the academic study of literature and the academic teaching of composition in a unified model of literacy, one that makes the fundamental distinction between texts and speech and not between, as is often the case, literary and nonliterary writing. The English profession, in other words, has been unable to draw strength from its historical origin in a unified model of literacy; it has been unable to integrate its practice in studying and teaching reading and writing with a unified theory of literacy and in so doing has consistently failed to recognize the symbolic nature of all texts, whether written by poets or historians, novelists or scientists, professors or students.

This failure to perceive the unified nature of literacy in part may be blamed on the persistence and flexibility of a long-standing rhetorical tradition, yet the persistence of rhetoric itself may be only a symptom of a deeper psychic strain within nineteenth- and twentieth-century cultural life. Why should the English profession, from its mid-nineteenth-century origins, have been so intent upon accepting the deeply personal, unifying power of literacy only as it relates to our reading experience and not to what we write and then only to the reading of certain kinds of texts labeled "literature"? The answer to this question may lie in the recognition that such an approach restricts the power of symbolic knowledge to a very narrow range of experiences—the reading and, for a few people, the writing of literary texts. Of all possible texts, only those classified as "literature" can be considered valuable, even though they are not subject to scientific verification—all other texts, including the writings of the teachers and students themselves, are to be subjected to narrow and readily accountable standards as to just what is or is not true.

English as an academic discipline, like many other aspects of nineteenth-century intellectual life, has always been insecure with its own understanding of truth, has always been too eager to be accepted by those disciplines whose concept of truth is grounded in the methodology of the natural sciences, and has thus always been too deferential to the claims of positivism. Yielding to these historical pressures, English professors have tended to see the reading of literary texts either as a source of a special knowledge unique to a narrow range of aesthetic experience (in some cases as narrow as the reading of lyric poetry) or as a healthy diversion from our relentless search for explanations of all experience, cultural as well as natural, that will attain the universality, the objectivity, and the ahistoricity of the physical sciences.

English professors, in other words, are accustomed to seeing the reading of literature as at once the source of all truth and the source of no truth. Literature tells us everything about the world, and at the same time it tells us nothing. The systematic study of a certain class of texts does yield truth, but it is a poetic, nonreferential truth that has no immediate connection with the real world. The one path that the English profession has never consistently followed, lacking as it does a unified model of literacy, starts with the assertion that the reading of literature is, not a distinct mode of experience yielding a unique kind of knowledge, but instead merely a *model* of the profound, imaginative, historical understanding of ourselves that we can attain through reading and writing generally. Such a tack, among other things, would establish historical and literary knowledge on a common ground and in so doing would threaten the long-standing accommodation that English and other academic disciplines concerned with human experience have maintained with positivism.

It is tempting, only a century after its origin, to dismiss academic English as an entrenched custodian of a narrowly defined literary tradition and to look for other sources to carry on the work of promoting literacy education in America. The difficulty with such a tactic, however, is that there is no other likely candidate. No other discipline has until now shown nearly as great an interest in the problematics of reading and writing texts. English does owe its meteoric rise in the liberal arts curriculum to its close historical connection with the emergence of a new model of literacy, while few other disciplines concerned with language growth and education generally afford any special status whatsoever to texts. For all its past shortcomings and despite many ongoing challenges to the problematic model of literacy from within the profession (to be considered in chapter 5), academic English does retain its general commitment to the notion that reading and writing entail special problems in the creation and comprehension of meaning.

The future of this commitment, however, may well depend on the ability of the profession to recognize just how short and just how tenuous are the threads that today connect us with the past. In particular, we need to recognize that what seems commonsensical today—namely, that reading entails comprehension and that writing entails creation—had little theoretical standing in educational philosophy before the rise of Romantic poetics and hermeneutics less than two centuries ago and had little practical impact upon curriculum and pedagogy before the educational reforms at the beginning of this century. What is so recent should not be mistaken for a given of our cultural existence; what has been in practice for only a hundred years need not remain in practice for another hundred. We must not misread the lesson of the past: those aspects of modern literacy education worth preserving will likely require the most vigorous defense and the most attentive nurturing.

Part Two
Practice

3

The Role of Praxis

I. Plato occupies an ambivalent position in the history of literacy because of his seemingly opposite reactions to the changes in Greek life associated with the emergence of writing. In the *Phaedrus*, Plato expresses grave reservations about the effects of what was then the fairly recent technological innovation of phonetic transcription. In this dialogue, Plato pictures writing as a threat to the humanizing aspect of ordinary human discourse. What is written down exists independent of any human agency; since the other is not present, the meaning of what is being intended is apt to be misunderstood. Writing also robs memory of its special value; what is inscribed will continue to exist if no one considers it worth remembering. Taken together, these arguments support the position that writing is derivative, not just phylogenetically and ontogenetically, in coming after speech in the course of human development, but ontologically as well, in being no more than a representation or picture of the world of speech.

The Plato of the *Phaedrus* is thus a progenitor of both the prelapsarian mythologizing of Rousseau and the structural linguistics of Saussure. He is a spokesman for the tradition that considers speaking to be the natural or preferred state of human communication, and whatever limited advantages writing offers come at a great cost. As Walter Ong notes in *Orality and Literacy*, hearing is the one sense that reveals knowledge about the inside of things—as when we tap a fruit to test by the sound for ripeness—and at the same time it is the sense that most completely engages the whole body. Drawing upon the work of Mircea Eliade, Ong refers to the omnipresent "centering action of sound": "For oral cultures, the cosmos is an ongoing event with man at its center. Man is the *umbilicus mundi*, the navel of the world" (1982:73). Sight, by contrast, operates by dissecting and breaking things up: "By removing words from the world of sound . . . print encouraged human beings to think of their own interior conscious and unconscious resources as more and more thing-like, impersonal and religiously neutral" (pp. 131–32). The basic contrast for Ong, as for the tradition 67 of the *Phaedrus* generally, is between an anterior natural world of sound and a posterior artificial world of sight.

Plato, however, also gives rise to a totally different tradition, one that champions literacy as the triumph of philosophy. According to classicist Eric Havelock, Plato's banishment of poets from the Republic is an attack not on practitioners of a certain type of literature but instead on people who wish to continue an outmoded, oral means of preserving culture. In a preliterate world, Havelock argues, verbal knowledge that had to be transmitted from one generation to the next was stored in the mnemonic rhythms of oral poetry, and individuals, in memorizing and reciting these verses, were espousing the worldview embodied in them in much the same uncritical and unphilosophical spirit that characterizes religious incantation.

According to Havelock, the poetic state of mind, not poetry itself, is Plato's "arch-enemy" (1963:47). The very interiority of sound, which in the *Phaedrus* Plato had seen as promoting a vital sense of harmony between doing and knowing, is now the object of suspicion. Individuals, Plato is warning, must now adopt an entirely new psychological stance toward knowledge—one that is critical and reflective instead of participatory and accepting. As Havelock writes, in banning poets Plato is "entering the lists against centuries of habitation in rhythmic memorised experience. He asks of men that instead they should examine this experience and rearrange it, that they should think about what they say, instead of just saying it. And they should separate themselves from it instead of identifying with it" (p. 47). In the *Republic*, Plato elevates symbolic exchange over synecdochic exchange and thus places literacy above orality.

Although the *Phaedrus* and the *Republic* discuss writing and oral poetry, their real concern seems to be different modes of transmitting knowledge or even different modes of thought. The primacy of speech advocated in the *Phaedrus* refers us to a world of tradition where knowledge and practice seem fully integrated, in which ideas and skills exist not divorced from one another but only as part of the unified social life of the community and where there is no knowledge independent of the collective practices of the group. In such a world, as cross-cultural psychologists John Gay and Michael Cole note, any one skill such as growing food is, not an "isolated technical activity," but a part in the total life of the community: "What Western cultures would compartmentalize into technical science, the Kpelle culture weaves into the whole fabric of existence. The relevant question is not 'How do you grow rice?' but 'How do you live?' " (1967:21).

Plato's point in the *Republic* is that certain patterns of traditional activity such as the memorization and recitation of poetic myths inhibit our ability to objectify what we know about practical activities like rice growing, inhibit our ability to see it for what it really is—not just a part of a series of communal traditions, including such things as chants, religious offerings, and ceremonial dress, whose collective practice leads to the annual production of the group's food supply, but also a technological means of raising food. Readers, unlike reciters, Havelock argues, are likely to "become the 'subject' who stands apart from the 'object' and reconsiders it and analyzes it and evaluates it, instead of just 'imitating' it" (1963:47). Readers, in other words, are likely to see questions about the efficacy of any single component in rice growing less as a threat to communal solidarity and more as a contribution to the well-being of the group by promoting greater agricultural yield.

Anthropologist Robin Horton develops this contrast in modes of thought by focusing attention on the central question of the presence or absence of a "developed awareness of alternatives to the established body of theoretical tenets" (1970:153). When knowing and doing are fully integrated, there is likely to be an absence of any such awareness, resulting in what Horton calls the characteristic *closed* quality of traditional life: "Absence of any awareness of alternatives makes for an absolute acceptance of the established theoretical tenets, and removes any possibility of questioning them. In these circumstances, the established tenets invest the believer with a compelling force. It is this force which we refer to when we talk of such tenets as sacred" (p. 154).

The *closed* quality of "traditional" life stands in opposition to the *open* quality of "modern" life, and Horton summarizes the fundamental differences between the "two basic predicaments" as follows: "The 'closed'—sacredness of beliefs, and anxiety about threats to them; and the 'open'—characterized by awareness of alternatives, diminished sacredness of beliefs, and diminished anxiety about threats to them" (1970:155). This "closed" mode of thought is characteristic of a world without symbolic meaning—there are no legitimate thoughts or actions apart from the prevailing social context. As Horton explains, "Despite their cognitive preoccupations, most African religious systems are powerfully influenced by what are commonly called 'emotional needs'—i.e., needs for certain kinds of personal relationship" (p. 161). Like the world of *Phaedrus*, Horton's "closed" world sees change only as decline and degeneration, corrected only by some cleansing, restorative action. The "open" predicament, on the other hand, like Plato's *Republic*, welcomes change as progress; in Horton's words, it promotes "the strongest possible incentive for a constant readiness to expose oneself to the strange and the disturbing, to scrap current frameworks of ideas and to cast about for replacements" (p. 170).

Between the "open" world of the *Republic* and the "closed" world of the *Phaedrus*, modern literacy scholarship has lodged its fundamental insight—namely, that the cultural shift from one world to the other, whether one views it as progress or decline, is intimately connected to how people communicate, specifically, to the presence or absence of writing. Modern work in the cultural origins of literacy has been erected upon the basic insight that *how* we communicate not only determines the form of our messages but ultimately affects their content as well and in so doing dramatically affects how we live and how we think. The transition from a traditional, "closed" world to a modern, "open" one, in other words, can be explained in large measure as a shift in how people communicate. As Walter Ong states, "Shifts hitherto labeled as shifts from magic to science, or from the so-called 'prelogical' to the more and more 'rational' state of consciousness . . . can be more economically and cogently explained as shifts from orality to various stages of literacy" (1982:29).

Such an approach to resolving the dichotomy within Plato, given wide currency by the works of Marshall McLuhan (1962, 1964), seems to have the advantage of abandoning ideological debates about the comparative value of different types of cultures in favor of dispassionate analysis of how the medium of communication shapes cultural life. As a reconsideration of Goody and Watt's seminal essay "The Consequences of Literacy" indicates, however, this position still leaves the most pressing questions in the Platonic dichotomy unresolved.

In their essay, Goody and Watt develop three principal points about the cultural origins of literacy. First, they claim that, in cultures without writing, verbal meanings are largely embedded in social action, or as they write, the "meaning of each word is ratified in a succession of concrete situations, accompanied by vocal inflexions and physical gestures, all of which combine to particularize both its specific denotation and its accepted connotative usages" (1963:306). Verbal knowledge results from the accumulation of such encounters, giving people a relationship with meaning that is "more immediately experienced by an individual" and is "more deeply social." Goody and Watt talk less about verbal knowledge than about memory, but their point is basically the same—that, in cultures without writing, memory plays a "social function" (p. 307). Only what is immediately valuable to the group and thus already embodied in the social context can be retained. As Goody explains in a later work, "Those innumerable mutations of culture that emerge in the ordinary course of verbal interaction are either adapted by the interacting group or they get eliminated in the process of transmission from one generation to the next" (1977:14). Something may start as an individual expression, but for the sake of its survival in a world without writing, its individual signature, to use Goody's metaphor, "tends to get rubbed out."

Second they argue that the critical moment in the transition from traditional to modern culture comes, not with the earliest forms of writing, such as runes or pictographs or even syllabaries, but only with

the development of phonetic transcription. Anticipating the more detailed work of Havelock, they make the point that all prephonetic forms
of writing were so crude that they could function only as mnemonic
aids, enabling people to inscribe and to read what they already knew.
Such prephonetic writing thus only tends to "reify the objects of the
natural and social order . . . [making] permanent the existing social and
ideological picture" (Goody and Watt 1963:314–15). Only with the
invention of phonetic transcription did people finally have a technology
capable of "expressing every nuance of individual thought," one that is
capable of expressing "personal reactions" as well as "items of major
social importance."

Finally, Goody and Watt argue that establishment of phonetic transcription is responsible for the separation of subject and object advocated by Plato in the *Republic* and, in turn, for the emergence of the
critical objectivity that has long been considered the defining characteristic of Greek and Western thought. Writing transforms the world.
No longer is verbal knowledge embodied in social relations; now, for
the first time in human history, knowledge can be codified and stored
in autonomous academic disciplines, independent of the beliefs of the
population at large. The study of the past thus becomes formalized as
history; the study of religion, theology; the study of public speaking,
rhetoric; and the study of thinking, philosophy. In creating distinct
areas of study, phonetic literacy also creates a new class of intellectuals,
people who specialize in one or more areas of knowledge not widely
shared by the group as a whole—a separation, Goody and Watt wryly
point out, that leads to the ongoing tradition of jokes about absent-
minded professors, the implication being that those who are better at
what the group cannot do must somehow be worse in what the group
ordinarily does as a matter of course.

Goody and Watt do their utmost not to read any developmental
pattern into their analysis of how literacy affects cultural life. In particular, they reject Lévy-Bruhl's formulation about the "prelogical" nature of traditional thought by citing anthropological studies that demonstrate its "logical" nature while simultaneously noting just how much
of Western thought and behavior is "illogical and mythical." They seek
to confirm neither the conservative mythology of the *Phaedrus*, with
its sense of a lost golden age, nor the progressive mythology of the
Republic, with its sense of the birth of a new age. Instead, they seek
to rise above controversy by explicitly relating the immense cultural
changes discussed by Lévy-Bruhl and others to the single question of
whether or not people are communicating in writing. In this central
paragraph they set forth a position that continues to be widely held:

Nevertheless, although we must reject any dichotomy based upon
the assumption of radical differences between the mental attributes of

literate and non-literate peoples, and accept the view that previous formulations of the distinction were based on faulty premises and inadequate evidence, there may still exist general differences between literate and non-literate societies somewhat along the lines suggested by Lévy-Bruhl. One reason for their existence, for instance, may be what has been described above: the fact that writing establishes a different kind of relationship between the word and its referent, a relationship that is more general and more abstract, and less closely connected with the particularities of person, place and time, than obtains in oral communication. There is certainly a good deal to substantiate this distinction in what we know of early Greek thought. To take, for instance, the categories of Cassirer and Werner Jaeger [namely, the fundamental modern distinction between myth and philosophy], it is surely significant that it was only in the days of the first widespread alphabetic culture that the idea or "logic"—of an immutable and impersonal mode of discourse—appears to have arisen; and it was also only then that the sense of the human past as an objective reality was formally developed, a process in which the distinction between "myth" and "history" took on decisive importance. [1963:321]

Goody and Watt do not deny the striking differences that exist between traditional and modern societies—such differences do exist, and Goody and Watt are quite ready to admit them—but their way out of this difficulty is to locate the source of these differences, not in genes or intelligence or even in social organization (all factors that may suggest a qualitative hierarchy), but in the medium of communication itself.

Such an approach does seem to defuse the potentially divisive issue of cultural bias and ethnocentricity. It is not important that people are different or even that they think and act differently, since the differences themselves are largely the result of whether or not phonetic writing is readily available. This type of explanation has been applied to the research that Russian psychologist A. R. Luria conducted with Asian peasants in the early 1930s as they underwent the experience of modernization. The peasants, Luria reports, regularly had trouble responding "correctly" to verbal syllogisms that Luria presented to them. "In the Far North," went one such syllogism, "where there is snow, all bears are white. Novaya Zemlya is in the Far North, and there is snow there. What color are the bears there?" A typical response is "I don't know what color the bears are there, I never saw them" (1976:107).

It does seem to make little sense to explain such a response in terms of cognitive development, especially in light of the fact that, when pushed for a response, the illiterate peasants would often concede the examiner's point, for example: "To judge from the place, they should be white. You say there is a lot of snow there, but we have never been there!" (p. 111). With only a little prodding, the peasant can make the "correct" deduction, but given his experiences with language, quite apart from his innate intelligence, he sees little point in taking seriously a verbal meaning that is not confirmed by personal observation.

There does seem to be little reason to label the peasant's response "prelogical" when it is based on firsthand experience and "logical" when it is based on a text—the one peasant, after all, is capable of either response and, on the basis of past experiences with language, merely prefers one to the other. Thus in one sense Goody and Watt's thesis does seem compelling—the medium of communication really determines modes of thought, or as Ong notes in reference to Luria's research, "It takes only a moderate degree of literacy to make a tremendous difference in thought processes" (1982:50).

The difficulty with Ong and, by extension, with Goody and Watt, is in clarifying precisely what is meant by "literacy." Indeed, the entire relevance of much contemporary work in the relationship between culture and literacy hangs on this very issue. One might well expect that the unproblematic model of literacy is being mentioned with the assumption that the cognitive and cultural fruits of literacy, as Goody and Watt themselves suggest, result from the power of phonetic transcription to give permanence to individually derived meanings. Yet a moment of reflection reveals that such an explanation is unduly reductionist. Clearly the presence of a system of phonetic transcription does not guarantee either collectively or individually the immense cultural and cognitive changes represented by what Horton describes as the "modern" or "open" predicament.

The mastery of phonetic coding skills, in other words, does not immediately transform our memories into history, our religious feelings into theology, or our thoughts generally into philosophy. When Goody in 1968 edited a volume of essays on literacy in traditional societies, it was quite apparent to him that the changes in Greek cultural life that in his earlier essay he had attributed to the development of phonetic transcription could no longer be described as the "consequences" of literacy, since none of the societies studied in the new volume of essays had followed the Greek model of cultural development after the introduction of a comparably flexible system of phonetic transcription.

Thus Goody was forced to admit that his original article should have discussed not the *consequences* but the *implications* of literacy and, in addition, that his real interest had always been in the "potentialities of literate communication" (1968a:4). But to speak of the "potential" of literacy is really to beg the question. What allows for the "full potentialities?" Why, we want to know, do not all societies that master phonetic transcription follow a pattern of development similar to that of ancient Greece? When we speak of the level of reading and writing attained in classical Greece as either a "consequence" or an "implication" of literacy or when we say that a certain mode of thought has been caused by literacy, we obscure rather than clarify key issues by supporting the notion that literacy itself is only a limited technical skill and not a complex cultural accomplishment.

In some sense the dialogues of Plato do represent the "consequences" or "implications" of literacy, but in another, more profound sense, they also represent its embodiment. We may assume, as observers have often done since Goody and Watt, that all cultures in the possession of transcription skills are literate, and we may then say that the creation of a substantial body of texts is the fulfillment of these skills. We can also assume, however, that a culture, like an individual, attains literacy only when it utilizes transcription skills for literate ends. At issue here is whether the study of literacy as a cultural phenomenon should focus first on its consequences or on its origins.

II.   The problem with considering literacy the source of significant cognitive change appears in the classification tasks that Luria gave the Russian peasants. In these tasks Luria presented the subjects with pictures of four common objects, such as a hammer, a saw, a hatchet, and a log, and then asked each subject to form a group that would exclude one item. For most peasants, customary usage had given the four objects strong synecdochic bonds, making it very difficult to form a new category, such as "tool," that would exclude the log—these tools, after all, are used in conjunction with logs. One sixty-two-year-old illiterate peasant, for example, was unable to see any reason whatsoever for excluding "wheel" from the series "knife," "saw," and "hammer." After Luria explained the possible use of the new category "tool," he gave the peasant another series—"bayonet," "rifle," "sword," and "knife,"—and posed the same problem. The dialogue then continued:

"There's nothing you can leave out here! The bayonet is part of the gun. A man's got to wear the dagger on his left side and the rifle on the other."

The principle of classification is explained: three of the objects can be used to cut but the rifle cannot.

"It'll shoot from a distance, but up close it can also cut."

He is then given the series *finger-mouth-ear-eye* and told that three objects are found on the head, the fourth on the body.

"You say the finger isn't needed here. But if a fellow is missing an ear, he can't hear. All these things are needed, they all fit in. If a man's missing a finger, he can't do a thing, not even move a bed."

Principle is explained once again.

"No, that's not true, you can't do it that way. You have to keep all these things together." [1976:60]

Such a dialogue does illustrate Ong's point that the peasant's thinking here is situational, not abstract; the peasant's "problem" by our standards is a seeming unwillingness to generalize, to consider groupings for what might be called symbolic classifications, not already embedded in practice. The central problem, in other words, seems to involve neither language use nor cognition exclusively but instead the general process by which the two activities are related; in both cases the subjects seem unable or at least unwilling to form groupings by rearranging items according to possibilities that are not found in customary use.

One major problem with determining the exact nature of the connection between being able to code and being able to group is the fact that most people learn coding skills in schools, where they also learn the principles of symbolic classification, thus making it very difficult to isolate the impact of learning to write from the overall impact of schooling. For years cross-cultural researchers Michael Cole and Sylvia Scribner suspected that proponents of the view that literacy transformed thinking were actually measuring not so much the effect of reading and writing as the impact of the system of formal schooling through which literacy was transmitted. Schooling, they believed, is an especially important factor, since by its very nature it removes students from the world of customary practice and places them in an artificial, decontextualized setting that increases the symbolic quality of all subsequent learning. In particular, schooling does seem to develop the ability to classify by logical relationships of form and function as well as by common practice and appearance.

Since schooling exerts a great influence on cognition, the researchers recognized the need to study the effect of literacy on a group of people with an indigenous written language, taught not at school but at home as part of the normal socialization process. Their goal, quite simply, was to find people who differed from others around them solely in that they could read and write—and then to study the cognitive differences between the two groups. Cole and Scribner did find such a people in the Vai, a tribe in Liberia that developed a written form of its native language to use mainly for record keeping and letter writing, activities that are not as well served either by English, the official language of the schools, or by Arabic, the language of religious instruction.

Two classification tasks as well as two tests of syllogistic reasoning are some of the many experiments that Scribner and Cole describe in the comprehensive account of their research, *The Psychology of Literacy*. The first classification task involved having the subjects sort and resort six cards that contained pictures of geometric shapes that differed in shape (circle or triangle), in number (one or two), and in hue (colored or uncolored). The test groups included nonliterate men, nonliterate women, monoliterates in Vai, monoliterates in Arabic, biliterates in Arabic and Vai (all groups that had not been exposed to formal schooling), and two groups of English literates divided by their number of years of schooling. The results were surprisingly uniform. All groups were able to form a single grouping, and all groups experienced difficulty with the task of reclassifying. The only dramatic difference was in the ability of those with ten or more years of schooling to offer a verbal explanation for their grouping, although they did not reclassify any better than the others.

Scribner and Cole also performed classification tasks modeled after Luria's, which required the subjects both to select a third item that belonged with two others (called "constrained classification") and to group twenty items belonging to the four common functional categories of food, clothing, tools, and utensils (called "free classification"). The constrained classification also used common objects from the categories of food, farm implements, and kitchen utensils. Again, the results indicated that coding ability did not have a significant effect on any group's ability to perform these tasks, nor, interestingly enough, did schooling—a result that countered the researchers' own expectations. Scribner and Cole note as an explanation that, unlike most cross-cultural cognitive studies, this one worked exclusively with adults whose current life-style did not necessarily reflect the fact that they had attended school even in the case of respondents who had attended school for ten or more years. The life-styles of the school-trained English literates in these studies were in fact generally the same as those of other groups tested. For these classification tests, the most important variable, therefore, was neither coding (as the Goody-Watt thesis would indicate) nor schooling (as Scribner and Cole expected it would be) but the life-style of the subjects, with those living in cities rather than in the country doing markedly better.

The results from the test of the ability to respond correctly to verbal syllogisms established a similar pattern but with some significant variations. In the first experiment, half the syllogisms supported current beliefs and half violated them, forcing the subjects to suspend disbelief and to accept the logical premise in order to answer the question correctly. On this task, neither Vai literates nor Arabic literates nor even those literate in both languages did any better than nonliterate subjects. Although schooling was a positive factor (more schooling resulted in higher performance), life-style continued to be significant: those living apart from modern Liberian life (for example, farmers) did measurably worse, regardless of any other factor. Thus it seems that the obstacle to be overcome in producing the consequences that Goody and Watt first suggested is not the inability to code language or even to think logically but the synecdochic force of traditional life that causes one to interpret experience generally as a necessary part of a common whole.

When Scribner and Cole, in their second test with a syllogism, constructed paradigms that had no basis whatsoever in ordinary experience (these syllogisms were about men and stones on the moon), they discovered that all groups did equally well. Throughout most of their tests, coding proved to be a very weak causal factor, and although schooling was important in leading to the breakdown in synecdochic responses, it seems to have been less so than the experience of the testing process itself: that is, all groups did significantly better when the test with syllogisms followed a whole series of tests—when, in other words, the subjects had become more fully adjusted to the demands of the new life-style represented by the testing process. Overall, the research of Scribner and Cole is immensely valuable in challenging the original Goody-Watt thesis. Scribner and Cole marshal considerable evidence in support of the notion that the introduction of a system of phonetic transcription is not a major determining factor in how people think. They do not consider, however, just what does lead people to use coding skills in the service of the literate activity of creating and comprehending texts.

Another sort of research study altogether approaches the issue of the cultural origin of literacy by considering a very different group of test subjects—American clerical workers and Ph.D. candidates. Both of these groups were unquestionably literate according to the unproblematic model, yet in the study conducted by Henry and Lila Gleitman they responded very differently to the task of negotiating symbolic verbal meanings. The Gleitmans required the participants to produce or to recognize suitable paraphrases of three-part phrases that are capable of producing two different meanings, depending upon the stress pattern; for example, the three-part phrase "black-bird-house" can become either a "BLACKbird house" (a house for a blackbird) or a "black BIRDhouse" (a birdhouse that is black). Using such combinations of simple words and stress patterns, however, also produces anomalous phrases like "BIRD-black house" (presumably a house painted a certain birdlike shade of black) and like the almost meaningless "bird HOUSE-black" (described by one enterprising Ph.D. candidate as "a blackener of houses who is a bird").

The Gleitmans screened both groups to ensure that all subjects could distinguish meanings according to stress patterns when the phrase resulted in two different but readily recognizable references (as in the case of "BLACKbird house" and "black BIRDhouse"); however, the responses of the two groups were almost totally different ("with no overlap at all") when the series of phrases was limited to items that resulted in anomalous or metaphorical meanings—meanings that violated synecdochic expectations. Thus "BIRDhouse black" was invariably interpreted as "black BIRDhouse" by the clerical workers as they tried to relate the phrase to their preexisting sense of reality. The Ph.D. candidates, on the other hand, just as consistently accepted what was given as a valid description of an as yet unfamiliar world. The phrase "eat HOUSEbird," for example, became a "house-bird who is very eat."

To overcome what the Gleitmans described as "massive differences . . . in the ability to perform a variety of tasks related to paraphrasing compounds," they attempted to reeducate the clerks by working through the sample lists, in their words, "over and over again, with feed-back as to correct choices and a financial reward for each correct choice made." Yet on retesting, the original disparity consistently reappeared, finally leading the Gleitmans to conclude that they "could find no simple means to teach the clerical group to perform as the Ph.D. group had" (1979:108). The research seems to have reached the conclusion that the two groups, despite the mastery of a high level of coding skills in both cases, responded differently to symbolic meanings.

The Gleitmans' study suggests that the key to the emergence of literacy seems to lie in the conditions that promote symbolic responses in actions as well as in words and thus that the limited but perhaps still important role of coding language must be placed within a larger picture relating language use to the general realm of symbolic activity. Just how such an investigation can proceed is indicated in a study conducted by Patricia Greenfield with Wolof children from Senegal. In this study, Greenfield divided children into groups on the basis of age and whether or not they were attending school. She then had the children perform simple classification tasks to see which groups were most likely to produce what she calls "superordinate groupings," or classifications based upon some general principle (such as color or shape) rather than upon a strictly personal, idiosyncratic criterion. Greenfield discovered not just that the schooled children were much better than the unschooled ones in forming superordinate groupings, or what we might call "symbolic classifications," but that within any one group the language used by the child to justify his or her response was a very powerful indicator of the child's performance.

Greenfield divided the children's verbal responses into three categories, ranging from holophrastic responses (for example, "red"), which depend most upon context, to assertions without the copula ("They red"), which specify more of the meaning in language independent of context, and ending with complete sentences ("They are red"), the verbal response that is most independent of context. Working with these categories, Greenfield discovered that the likelihood that a superordinate grouping would be formed increased threefold for the children attending school and sixfold for the children who were not attending when the verbal response took the form of a complete sentence with the copula. These findings suggest that the ability to perform symbolic action is connected with the ability to negotiate symbolic meaning—and in both cases, as Greenfield herself asserts in this general comment, the key factor seems to be the individual's freedom to break through the constraints of customary practice:

> The embedding of a label in a total sentence structure (complete linguistic predication) indicates that it is less tied to ια, situational context and more related to its linguistic context. The implications of this fact for manipulability are great: linguistic concepts can be turned upside down more easily than real ones. Once thought is freed from the concrete situation, the way is clear for symbolic manipulation . . . in which the real becomes but a sub-set of the possible. [1972:174–75]

The value of Greenfield's insight here is that it points us in the direction of seeking not a causal but a developmental relationship between action and language use, and specifically between the ability to perform certain actions demonstrating symbolic thought processes and the ability to generate and, most likely, comprehend symbolic language as well. In Greenfield's study, as in the Gleitmans' and Luria's, the basic problem for all groups was the inability or reluctance to cast aside expectations embodied in customary behavior and attitudes. The presence or absence of coding skills thus seems to have little directly to do with impeding or promoting within each group the very skill that lies at the heart of literacy—namely the development of the ability to respond symbolically to anomalous experiences, the development of the ability to treat as valuable in themselves those experiences that have no clear and obvious connection with ordinary social existence. We therefore need to investigate this developmental relationship between thought and action, for only in the larger account of the growth in general symbolic activity can we trace the origins of literacy itself.

III. In *Play, Dreams, and Imitation in Childhood* (1962), Jean Piaget traces the development of representational activity in the interaction of imitation and play. Imitation for Piaget represents the domination of the principle of *accommodation*—wherein we alter ourselves so that we conform to the demands of the world. Similarly, play represents the domination of the principle of *assimilation*—wherein we attempt to change the world either in thought or in reality to conform to our wishes. Cognitive growth throughout childhood, according to Piaget, follows a dual pattern of development. On the one hand, there is the effort to maintain an equilibrium between accommodation and assimilation; without this equilibrium we alternately experience as dominant first representative imitation (as we mimic the world around us) and then symbolic play (as the world changes fancifully according to our whims). On the other hand, equilibrium by itself is insufficient for cognitive growth, since both representative imitation and symbolic play are individually in need of development.

For cognitive growth to occur, therefore, both imitation and play must become increasingly complex and transposed as our imitative and playful responses come to depend less upon the original stimulus that triggered them. Our impetus toward imitation thus develops in the direction of becoming increasingly "deliberate," that is, freely chosen by us and not controlled by whatever happens to be before us at the moment. Unlike infants we are free to select our own models of behavior. Meanwhile, our impetus toward play develops in the direction of becoming increasingly "constructive," that is, exhibiting a structural complexity that reflects real rather than fanciful transformative powers. Again, unlike infants we really can transform the world, turning clay into pots. Piaget, in a key passage, describes this basic pattern of human development as follows:

Gradually as the differentiation and coordination of assimilation and accommodation occur, experimental and accommodative activity penetrates to the interior of things, while assimilatory activity becomes enriched and organized. Hence there is a progressive formulation of relationships between zones that are increasingly deep and removed from reality and the increasingly intimate operations of personal activity. Intelligence thus begins neither with knowledge of the self nor of things as such but with knowledge of their interaction, and it is by orienting itself simultaneously toward the two poles of that interaction that intelligence organizes the world by organizing itself. [1954:354–55]

Equilibrium itself is not the central issue, since all forms of life, including that of the simplest organisms, attain some balance between the demands of assimilation and accommodation; what is essential for human cognitive development is the progressive deepening of both assimilation and accommodation. Our representational powers increase, Piaget contends, only as we seek an equilibrium between a mode of accommodation that is increasingly deliberate and a mode of assimilation that is increasingly constructive. Unlike other animals, human beings seek an equilibrium based upon ever-greater control of the natural world and ever-greater awareness of the limits of that control.

The schematic explanation of cognitive growth offered above is most helpful in explaining relative failure and success in the performance of a variety of symbolic activities. For example, the difficulty that Luria's peasants had in omitting one item from the group "hammer, saw, hatchet, and log" can be explained in terms of the domination of accommodation—the original four-item group is in fact an imitation of a naturally occurring group. The extent of the domination of accommodation is reflected in the unwillingness and perhaps even inability of the respondents to reshape the world; the grouping of "hammer," "saw," "hatchet," and "log" is so embedded in actual practice that it resists all efforts at assimilation. There seems to be no sense in which the respondents are accustomed to "playing" with such a grouping. The clerks in the Gleitmans' study, on the other hand, could not overcome the domination of assimilation. Collectively, they could find no reason to conform to the new reality presented by the anomalous phrase "BIRDhouse black," regardless of the inducement the Gleitmans offered; they preferred instead to transform the world before them, as if magically, into something familiar. Thus the "BIRDhouse black" inevitably became "BLACK birdhouse." The world given to them in the three-part phrase was rejected for their own sense of what should be.

The same forces of accommodation and assimilation at work in these two experiments affect cultural adaptation generally. For example, as we have seen, knowing and doing are highly integrated at the center of traditional cultures, and this fact can also be explained as reflecting the alternate domination of imitation and play. When the central question is not "How do you grow rice?" but "How do you live?" individuals necessarily experience far greater pressure to accommodate to the prevailing practice; indeed, there is no way to effect even a limited agricultural improvement without simultaneously challenging the order of the community. As Horton notes, group practice in such a situation takes on the status of a ritual, and the individual's principal obligation becomes to act in accord not with some pragmatic, scientific, or universal standards that might, for example, increase rice production or lead to more equal distribution but with the rules of the ritual. A true-false criterion, S. J. Tambiah adds, is often inappropriate where language use and action are embedded in ritual; in such situations, the criteria of acceptability are "better conveyed by notions such as 'validity,' 'correctness,' 'legitimacy,' and 'felicity' of ceremony performed" (1973:219).

Walter Ong goes even further in asserting that the most vital communication in what he calls "oral" cultures is generally a reworking of that which is already known, an elaboration of synecdochic meaning. In other words, communication in an oral culture is more likely to be considered "an invitation to participate, not simply a transfer of knowledge from a place where it was to where it was not" (1977:118), and hence even less the creation of a mode of being that is unsupported by synecdochic practice.

We cannot doubt the attraction inherent in such ritualized communication, especially for intellectuals who live much of their lives in a world of symbolic texts. Stanley Diamond, for example, in a sympathetic account of "primitive" society, refers to the ritual drama at the center of so much oral language as "a culturally comprehensive vehicle for group and individual expression at critical junctures in the social round or personal life cycles" (1963:97). The therapeutic value of such communication and other benefits, however, are often inseparable from the accommodative pressures they exert. J. C. Carothers, in a study of preliterate psychology, claims that language and action are often so well integrated in preliterate societies that all speech is apt to be viewed as a part of the behavior it accompanies and hence as a form of action. Thus to the extent that the language to be used in conjunction with the planting and harvesting of rice is an integral part of the whole process of rice growing and, by extension, of being a member of a particular community, then there will undoubtedly be pressure to ensure that such language is as imitative as the technology it accompanies. The problem with such a world is how to reconcile the benefits that Diamond and many others see in ritual with the generalized accommodative pressure of the sort related by Gay and Cole:

> In one case a teacher in a nearby school told a child that insects have eight legs. The child (who worked in the Gay household) one day happened to bring an insect to Mrs. Gay. They discussed the fact that it had six legs, contrary to the teacher's remark. The child, with Mrs. Gay's encouragement, took the insect to school to show the teacher. The child was beaten for his effort—and insects continued to have "eight legs." From the teacher's point of view, the important thing to learn was a set of words and respect for authority. [1967:33]

Here the pressure for verbal accommodation is expressed as part of a larger cultural need for imitative behavior. As Ong notes, the constant emphasis on mnemonic activity in certain societies does "more than develop memory": "It creates what can be called a mimetic culture, a culture of imitation, a state of mind that values copying" (1977:285). What is not as clear from such an explanation, however, is the extent to which the domination of the accommodative pressure to conform is in turn based upon a comparable assimilative pressure to transform the world as if magically. For the child in this anecdote, as for the doctoral candidates in the Gleitmans' study, a given state of affairs, even if it violated expectations based upon a teacher's authority or even if it violated almost all bounds of verbal usage, nonetheless still required an accommodative response from the individual. For the teacher in this anecdote, however, as for the clerks in the Gleitmans' study, the world is much more malleable, seeming to conform to their wishes in what Piaget calls the ludic spirit of play. The accommodative demands for strict imitation are merely an extension of the domination of the teacher's unconstructive assimilation.

Anthropologist C. R. Hallpike (1979), in his Piagetian study of primitive thought, reveals that accommodation and assimilation are interconnected by comparing the process of housebuilding in modern and traditional societies. In modern housebuilding, nature is substantially assimilated to conform to our desires with respect to the house's appearance and comforts. These assimilations are, to use Piaget's key term, fully "constructive." Meanwhile, all our choices about location, plans, materials, and costs reveal not only our constructive assimilative power but also the many practical limits to what we can afford to build. Transforming nature in the process of housebuilding is thus only partially an act of assimilation—we must also fit our plans to an image or model of what is technically and financially possible. This accommodative act of self-control is fully developed only if, to use Piaget's term, it is "deliberate," that is, only if it represents a mature reconciliation of the desirable and the possible. In building a house, our technological mastery of nature and our constant need to accept limits go together, creating an equilibrium of *constructive assimilation* and *deliberate accommodation*—we must learn to live with almost unlimited choices and unlimited constraints, and we can proceed only by constantly reconciling the two.

People who live without such technological control over nature, Hallpike argues, have fewer choices (which seems obvious) and fewer constraints as well. The housebuilder often does the work himself with materials that are available in abundance. The size, arrangement, and construction of the house directly reflect the builder's everyday experience, and the resulting structure partly transforms and partly imitates nature. The thatching, for example, in not losing its distinctive properties, only partly assimilates nature, yet in retaining some semblance of its original function, it is also partly imitating its original function. The assimilative changes that the builder works on nature are thus more readily accomplished and meet less resistance than those encountered by a modern builder, because they are much less complete and, in turn, much less constructive.

The builder who works from a tradition that guarantees success as long as he stays within it encounters few of the obstacles and clearcut decisions that time and again reveal to his modern counterpart the definite practical limits of his own endeavor. The traditional housebuilder, in other words, senses fewer obstacles before him because customary practice leads him to choose the well-trodden path. At the same time, the house that eventually emerges is more closely integrated with the builder's life, in large measure because it is neither freely chosen nor fully transformed.

The ready assimilation of nature and the prevalence of strict patterns of accommodation are not opposites but corollaries, each expressive of the more encompassing equilibrium by which a group maintains itself between the transforming power of its technology and the intractability of nature. All peoples confront this problem: constructive assimilative acts must transform the natural world before the material basis of life can be provided, yet given the limits of any technology, we must also accommodate ourselves to conform with the limits of that transformed world. As a rule, when we have less power actually to transform the world, less power to effect constructive assimilations, we tend to experience both greater ease of assimilation and greater pressure to accommodate. A limited technology is accompanied by greater social pressures prohibiting anomalous thought and a greater sense of the efficacy of playful assimilations or magic. The very ease of assimilation tends to mask the accommodative pressures to conform, creating at once fewer options and (seemingly) fewer restraints.

IV. As the volume of essays *Literacy in Traditional Societies* well illustrates, the attempt to explain cultural change in terms of the introduction of writing succeeds only in begging the larger question "What promotes literacy?" There is no guarantee that the introduction of coding skills will radically transform any society; a society seems just as likely instead to incorporate those features of coding that best serve its current technological level of development. The Targu in Melanesia, for example, have incorporated writing into their chiliastic cargo cult and consequently see the writing of letters as a form of incantation in which contact is established with a supernatural world rich in material rewards—letters are sent overseas and ships then arrive filled with cargo. Similarly, the natives of Madagascar use literacy mainly to read books on astrology, with those in French being especially popular, their strangeness giving them added oracular power. Indeed, apart from people's practical interaction with the world through activities such as trade and public administration, coding skills almost everywhere seem to serve mainly a narrow religious purpose, with "writing" consisting of little more than the transcription of sacred manuscripts and "reading" amounting to little more than the public recitation of such works.

Melanesia and Madagascar are no more the "consequence" of literacy than is classical Greece. People flourish, and in the process make use of coding techniques and other technological innovations, only as part of a general historical effort to deepen the existing equilibrium between accommodative and assimilative practice—that is, only within the context of true cultural development.

Literacy therefore entails the use of transcription or coding skills within the context of a larger pattern of human development in which play becomes increasingly constructive and imitation becomes increasingly deliberate. Transcription is a technological improvement over human memory, but literacy emerges only when individuals utilize the technology in the process of deepening their own level of symbolic representation. The use of coding skills within a larger pattern of growth thus best accounts for the value of literacy and the cognitive growth that is sometimes associated with it. Here Scribner and Cole correctly reject the notion that somehow the coding skills themselves have some sort of magical transforming power apart from the larger life of the individual. In addition it is possible to understand better why they reasoned that schooling was apt to play a far greater role in cognitive life than the mastery of coding skills, since so much of schooling is purposefully designed to promote a deepening of accommodation and assimilation.

In an earlier study, for example, Cole and Scribner do attribute three fundamental cognitive changes to schooling, each of which is sometimes associated with literacy. First, "attendance at school apparently encourages an approach to classification tasks that incorporates a search for a rule—for a principle that can generate answers" (1974:122). Second, and even more important, "schooling seems to promote an awareness of the fact that alternative rules are possible—one might call this a formal approach to the task in which the individual searches for and selects from the several possibilities a rule of solution." Finally and, as they note, unambiguously, "schooling (and only schooling) contributes to the way in which people describe and explain their own mental operations." A mastery of coding skills, they hypothesized and then demonstrated among the Vai, should not have comparable general cognitive effects.

Cognition develops, not as a result of literacy, but in response to practical interaction with the world. The mastery of coding skills should therefore produce cognitive changes only in those areas specifically related to the actual use of these skills. When nonschooled, literate Vai did not outperform their nonliterate neighbors in standard psychological tests involving classification and logical deduction, Scribner and Cole attempted to prove the other half of their hypothesis—namely that literate Vai should do better on tests measuring some of the specific skills that are ordinarily mastered by the Vai in the course of learning how to read and write outside of school.

One such test measured the ability of literate and nonliterate Vai to give clear, informative directions, a skill that Scribner and Cole reasoned is necessarily fostered in the course of letter writing. To test for this skill, they devised two specific tests—one asking various groups to explain the formal play of a simple game and the other asking them what was assumed to be a more indigenous and hence more indicative task, namely, to give both strangers and their own wives directions to their farms. The results generally supported their premise that the ability of literacy to enhance performance on a variety of skills depends largely upon the nature of the reading and writing tasks involved. For example, Vai literates started half of their responses with some general orienting information (such as, "I am now going to explain to you how to play a game"), while only 20 percent of the nonliterates did so. With the support of these results, Scribner and Cole attempt to formulate a definition of literacy in terms of one's ability to perform specific practical activities: "Literacy is not simply knowing how to read and write a particular script but applying this knowledge for specific purposes in specific contexts of use. The nature of these practices, including, of course, their technological aspects, will determine the kinds of skills ('consequences') associated with literacy" (1981:236).

In moving beyond the narrow range of the unproblematic model, this definition of literacy is unquestionably correct. Practical activity does have as much to do with literacy as the mastery of coding, if not more. The difficulty with Scribner and Cole's definition of literacy, however, is that it lacks any developmental component. They conceive of a world of diverse practical activities, some of which involve reading and writing while others do not, but with each one having its own special value to either the individual or society. The key term "practice" Scribner and Cole define as "a recurrent, goal-directed sequence of activities using a particular technology and particular systems of knowledge" (1981:236). Individuals, they contend, are engaged in "social practice" whenever they use a "shared technology and knowledge system" to perform tasks in the pursuit of "socially directed goals."

In Scribner and Cole's formulation, the conditions under which the interaction of knowledge, skill, and technology can accurately be labeled "goal directed" are unclear. In a weak sense, "goal directed" may mean only the purpose of an activity and in this sense may be applicable, for example, even to animal behavior. In a stronger, more problematic sense, however, an activity should be called goal directed only to the extent that the subject performing the activity is free to select one goal rather than another. In this sense, "goal directed" refers solely to human pursuits and even then only selectively. Scribner and Cole's notion of practice does speak to the need for the *constructive transformation of play*; that is, we are often building something quite intricate in practice, but it neglects any consideration of the extent to which this activity also represents the *deliberate transformation of imitation*, that is, the extent to which we are free to act one way and not another.

We can see the limits of such a notion of practice by considering a statement in Marx's *Capital* that Scribner and Cole use as an epigraph in their introduction to a 1978 collection of Vygotsky's essays:

> We pre-suppose labor in a form that stamps it as exclusively human. A spider conducts operations that resemble those of a weaver, and a bee puts to shame many an architect in the construction of her cells. But what distinguishes the worst architect from the best of bees is this, that the architect raises his structure in imagination before he erects it in reality. At the end of every labor process, we get a result that already existed in the imagination of the laborer at its commencement. He not only effects a change of form in the material on which he works, but he also realizes a purpose of his own that gives the law to his *modus operandi*, and to which he must subordinate his will. And his subordination is no mere momentary act. Besides the exertion of bodily organs, the process demands that, during the whole operation, the workman's will be steadily in consonance with his purpose. [1967a: I, 178]

The architect's purpose, however, is constrained not just by nature itself, as Marx suggests, but also by the depth of the equilibrium that has been established between accommodation and assimilation—the greater the power to transform nature, the less constrained and hence more deliberative the imitative act. We need to recognize the developmental pattern in the very act of setting goals, with one's freedom to select a course of action directly related to one's power of constructive assimilation. Accordingly, what Scribner and Cole call "practice," or goal-directed activity utilizing skills, knowledge, and technology, must be divided into two distinct classes—one class including all such activities that tend to preserve the existing balance between assimilation and accommodation and a second class including those activities that tend to create a new, weightier balance by making accommodation more deliberate and assimilation more constructive. The first such class, which involves the repeated performance of tasks necessary to maintain the existing balance between our needs and our technological limits, might well be called *practice*. The other class, which involves those activities that promote a new equilibrium between accommodation and assimilation whenever the balance maintained by current practice proves inadequate, would then be classified as *praxis*.

It is essential to recognize from the outset that practice and praxis are not to be differentiated on the basis of any intrinsic qualities; indeed, all forms of practice must originally have existed as acts of praxis. Furthermore, the complexity of any social practice has no inherent limits. The practice of any one group may be more compelling in almost every measurable way than the praxis for another group. In addition, as Durkheim (1933) observed, societies with limited powers of assimilation are likely to manifest far less division of labor, so that individuals must master a host of complex practices each of which is beyond the immediate grasp of even highly educated people in more technologically advanced societies. Praxis is to be preferred over practice (and here is the real source of confusion in disciplines concerned with the comparative study of culture), however, only in the sense that one particular activity represents an advance over a former activity *within a single context.*

What distinguishes praxis from practice therefore cannot be shown in any synchronic comparison between different activities. Instead, this difference can be seen only in a diachronic or historical evaluation of a single tradition and then only if it can be demonstrated that this one action represents a developmental advancement. Praxis, in other words, can be identified as only part of a developmental model reflecting the unfolding of an individual's or a group's history. Without movement toward some goal and therefore without some sense of progress, there can be no praxis.

Praxis, not coding, explains literacy, for literacy is ultimately best described, not as coded speech, but as verbal praxis; that is, literacy comes to life only in the efforts of people to use written language in acts of praxis. Such a definition of literacy is especially helpful in revealing the confusion in Scribner and Cole's effort to divorce schooling from "literacy." While they do succeed in revealing the bankruptcy of the unproblematic model, they do so at the cost of divorcing from our sense of literacy the alienating, symbolic, praxis-directed activity that is characteristic of much schooling and is absolutely essential for literacy. "The school," writes cognitive psychologist Jerome Bruner, "is a sharp departure from indigenous practice" (1965:1009). Merely placing learning in an unfamiliar context—the "extirpation" of schooling, in Bruner's words—renders it "an art in itself freed from the immediate ends of action, preparing the learner for the chain of reckoning remote from payoff that is needed for the formulation of complex ideas."

Scribner and Cole's research is thus misdirected to the extent that it seeks to reveal the limited effect of "literacy" by ensuring from the outset that "literacy" itself is limited. Their research project, in other words, precludes the possibility of their recognizing that reading and writing can become fully literate activities only as forms of praxis, that is, only when they are "freed from the immediate ends of action." As we learn at the end of their study, precisely this freedom has never developed for the Vai: "Vai script literacy," they conclude, "is not a vehicle for introducing new ways of life. We have called it literacy without education because it does not open doors to vicarious experience, new bodies of knowledge, or new ways [of] thinking about major life problems" (1981:238). The great cognitive effects of literacy, quite simply, are the result not of coding but of praxis. People who read and write without engaging in praxis do not show marked cognitive change, and there is no reason to believe that they should. The research project of Scribner and Cole is restricted from the outset to revealing the limited value in the *practice* of reading and writing. It has little directly to say about the more important question—the value of literacy as verbal praxis.

The circumstances surrounding the Asian peasants studied by Luria in the early 1930s, however, were markedly different. Luria recognized that the "basic categories of human mental life" are best understood as "products of social history—they are subject to change when the basic forms of social practice are altered" (1976:164). It is more important, not that these peasants learned how to read and write, but that they learned how as part of a larger effort that, at least at the time, seemed to offer them the prospect of wielding greater control over their own lives. The central element in the great cognitive changes that Ong attributes to "literacy" is not the mastery of coding skills but the creation of new motives:

> These complex motives, which go beyond concrete practical planning, assume the form of conscious planning of one's own labor; we begin to see interests that go beyond immediate impressions and the reproduction of concrete forms of practical activity. These motives include future planning, the interests of the collective, and, finally, a number of important cultural topics that are closely associated with the achievement of literacy and assimilation of theoretical knowledge. [Luria 1976:162]

These very complex motives are necessary if one is to resist the accommodative pressure to reproduce "concrete forms of practical activity," that is, to imitate the prevailing practice, and instead to engage in the kind of thinking, planning, and general activity that is based on new, imaginary, or playful forms of being.

To resist the pressure to copy what is, we must rely on an image of what might be, an image that may exist only as assimilative fancy within the individual but that, as it becomes more organized and more constructive, forms the basis of the world of theoretical knowledge. Living in a world that offers little sense of assimilative constraint and little sense of freedom from accommodative pressures, we can have only a very limited sense of a world of theoretical knowledge—without restraints there are no possibilities, and the practice that supports such an existence often passes unnoticed by us. Practice seems to be a part of our natural condition of being in the world, or to use the metaphorical language of Heidegger, our practice exists as a "given-to-hand," something so familiar that it goes unnoticed.

Such a state, however, does not necessarily continue indefinitely—circumstances change, by natural process, by coincidence, or by meddling. At some point, we "bump" against the world and for once notice the practice or given-to-hand as a "thing-out-there." No longer is the practice part of the natural world; no longer must we accommodate ourselves to it. Now, for the first time, it is seen for what it is—a possibility, one way of being in the world. With this shock of recognition comes the birth, first of theoretical knowledge, then, by extension, of praxis (that is, action directed by such knowledge) and of literacy (the verbal form of such action). At the foundation of literacy, therefore, is theoretical knowledge derived from our realization of the contingent nature of the world in which we live. What is *actual* we share directly with others based upon our common social experiences; it is our world of synecdochic knowledge. What is *theoretical* we share only indirectly with others through the service of mediating structures such as texts; it is the world of symbolic knowledge. When we act to transform theory into practice, we are engaged in praxis, and when that praxis utilizes writing, we are engaged in literacy.

4

The Ontogeny of Literacy

I. There is no more wondrous, insightful, and in key ways misleading account of the origins of language than Helen Keller's description of her initial encounter with words:

She brought me my hat, and I knew I was going out into the warm sunshine. This thought, if a wordless sensation may be called a thought, made me hop and skip with pleasure.

We walked down the path to the well-house, attracted by the fragrance of honeysuckle with which it was covered. Some one was drawing water and my teacher placed my hand under the spout. As the cool stream gushed over one hand she spelled into the other the word *water*, first slowly, then rapidly. I stood still, my whole attention fixed upon the motion of her fingers. Suddenly I felt a misty consciousness as of something forgotten—a thrill of returning thought; and somehow the mystery of language was revealed to me. I knew then that w-a-t-e-r meant the wonderful cool something that was flowing over my hand. That living word awakened my soul, gave it light, hope, joy, set it free! There were barriers still, it is true, but barriers that in time could be swept away.

I left the well-house eager to learn. Everything had a name, and each name gave birth to a new thought. As we returned to the house every object which I touched seemed to quiver with life. That was because I saw everything with the strange, new sight that had come to me. [1903:23–24]

As Piaget explains, there is a gradual developmental movement between such actions on the sensory-motor level and thinking based in representative images. The action of being handed a hat and the formation of an image or cluster of associations involving the hat and signifying the possibility of a walk are not necessarily distinct experiences waiting to be linked at the onset of language. Even before her encounter with language, Helen Keller was capable of recognizing that the simple act of being handed a hat signaled the likelihood of her going outside. Certain qualities of the hat rather quickly become asso-

91

ciated in the child's mind with the joy of going outside, in such a way that she is capable of using the hat as a sign in order to communicate at a preverbal level the likelihood of going for a walk or the desire to do so.

Yet Miss Keller is celebrating language, not just communication. This narrative supports the notion that a verbal apprehension of the world represents a qualitatively higher form of knowing, one that is uniquely human. The key moment in Helen Keller's development (and in human development generally) is the product not of preverbal signification or communication but of language. In terms of the model of development presented in the last chapter, the gesture with the hat requires only minimal acts of accommodation and assimilation. On one hand, such gesturing is clearly imitative, representing the reproduction of a significant portion of the desired original action; to the extent that the portion selected is the product of the compelling feature of the activity itself, then the accommodation is far from deliberate. On the other hand, the gesturing with the hat also represents a limited level of assimilation.

To the extent that the gesture is intended to provide only a minimal clue as a ready means of triggering a desired action, any efforts to enhance the structural complexity of the sign itself would be counterproductive. When we take an interest in the internal nature of the sign itself, as we must in interpreting metaphor, we delay the consummation of the communicative intent in our appropriate response.

In addition, by relying heavily upon key aspects of the original experience (as in the association of hat with walks), minimal acts of assimilation reinforce the notion that the sign itself has a real transformative power and not just a playful one. Thus it is possible for a person to believe that a sun hat really does in some way precipitate enjoyable walks, and in such situations it would seem to be natural for the sign itself, here the hat, to take on intrinsic, almost magical powers, becoming a fetish, while the enactment with the sign, here the imitative gesture, would be experienced as a primary cause of some subsequent behavior, and not just as an announcement.

The accommodative-assimilative component of such a limited act of signification as gesturing with a hat reveals the similarities and differences between the Piagetian developmental model of cognition and the categories of signification—icon, index, and sign—formulated by American philosopher C. S. Peirce (1932). An "icon," according to Peirce, is a representative segment or sample of whatever is being represented, as, for example, when a stick drawing is used to represent a human being. In imitating essential aspects of what they represent, icons are fundamentally accommodative.

An "index," conversely, is any element within a sequence of activity whose presence indicates or somehow seems to cause, and in so doing, refs to, some seemingly causally related condition, as in the sense that dark clouds are an index of rain. Although an index depends upon the existence of a causal connection between a sign and its referent, this connection is just as likely to be psychological as natural, that is, just as likely to be based upon expectations derived from common experience as upon an actual natural or mechanical connection. While the total confusion of natural and psychological realms perhaps characterizes magical practice and belief, the partial overlapping of these categories is an everyday experience. Signs that function as the hat did for the young Helen Keller frequently derive force and value from our sense of their active participation in an interrelated series of events. In controlling the sign, we often exercise some control over the subsequent events or at least sense that we do. The index is therefore a function of our efforts to assimilate the world, initially at least and most basically, through fanciful, unconstructive transformation.

While Peirce regards icons and indexes as two separate classes of signs, from a developmental perspective they are interrelated. As with action generally, the iconic force of a sign is directly related to our sense of its immediate transformative power. Were a young child to gesture with a hat, she would be expressing alternately the dominance of accommodation, to the extent that she was reconstructing an important element of the original experience, and the dominance of assimilation, to the extent that she sensed her actions would actually bring about the desired response as part of a likely chain of events. In both instances the referent of the gesture is determined by a pre-established structure of events. The iconic meaning of gesturing with the hat is thus established by the outstanding feature of the activity itself, just as the indexical meaning is established by the customary outcome.

Meanwhile the context is so well determined and so supportive of meaningful communication that only a minimal clue offered by one party is needed to trigger the appropriate response in the other. The purely verbal question about going for a walk, however, seems to break through the limit of iconic and indexical limitation and thus to be a necessary part of Bruner's category of "symbolic" cognition. The verbal request seems to be free of synecdochic expectation in having neither imitative nor causal connection with a given chain of action. Language thus generally seems to be a part of Peirce's third category, what he calls "signs" but what is perhaps better conveyed as "symbol"—that is, the class of signs that is based upon an arbitrary rather than a necessary connection with referents. Symbols are different from icons and indexes in that they lack both the physical resemblance to what they signify and the causal connection with any outcome.

The progression from either icon or index to symbol thus seems to follow a general developmental pattern in which meaning emerges from deep within specific contexts or is freed from dependence on any one context. Only the symbolic reference of language—what Helen Keller called the "living word"—transformed her consciousness. She did experience preverbal kinds of signification, what she called a "wordless sensation," but only the contact with the universal, symbolic power of language made her feel fully human. As philosopher Ernst Cassirer has noted, only the word is the kind of sign that "in contrast to the actual flow of the particular contents of consciousness has a definite ideal *meaning*. ... It is not, like the simple sensation, an isolated particular, occurring but once, but persists as the representative of a totality, as an aggregate of potential contents, besides which it stands as a first 'universal' " (1968:89). With a knowledge of the word "water," the young Helen Keller entered the world of symbolic reference, of possibility, and of theoretical knowledge. The passage itself, supported by Cassirer's observation, thus seems to support the claim that language itself forms the basis of distinctive human existence.

Unaddressed in this sequential account of the development of signs, and perhaps even obscured, is the relationship between the onset of language, described so eloquently here by Helen Keller, and the "barriers" she spoke of still having to overcome. Does the onset of language itself guarantee the fulfillment of the symbolic thought processes it encourages? Words certainly do have great potential to convey symbolic meaning; they are generally far freer from synecdochic ties than either icons or indexes. It is not equally clear, however, whether the symbolic power of language originates in the words themselves or in our ability to use them for symbolic purposes. Are we capable of symbolic expression because we are capable of speech, or is our capacity for speech part of a more generalized capacity for symbolic development? Are words always symbols, and must they therefore always be used symbolically? Language unquestionably has a special role to play in human development regardless of the answer to these questions, but we need to know what course of action to follow to attain not the initial use of language described by Helen Keller but the level of literacy that is evident in her writing.

II.   Problems about the symbolic nature of language begin with the notion of *words* as the most elemental unit of speech, that unit from which all larger units and hence speech generally is derived. Words seem to be inherently symbolic, lacking either the mimetic quality of icons or the causative quality of indexes. Since words are the atomic elements of sentences, it seems reasonable to conclude that language generally, as the combination of sentences, is in turn necessarily symbolic. Such premises and conclusions do serve as the basis for developmental schemata like Bruner's, in which thought is seen as evolving in a series of stages, from a sensorimotor to an imagistic to a verbal level (Bruner and Greenfield 1966).

What is symbolic about words, however, is not words themselves as they are regularly used in utterances but only the *concept of a word* as a distinct unit of language, either meaningful itself or meaningful in combination with other words. Indeed, while the linguistic phenomenon of words may be universal, an awareness of the concept of words as symbolic units of signification seems to be generally limited to those people with knowledge of writing. Few languages of the world without a written form have a word for *word*, that is, have a word by which to refer to the concept of discrete, meaningful verbal units that are combined to form utterances.

Scribner and Cole, for example, discuss the difficulty they had asking native Vai speakers to name the longest word they knew, since the Vai language has no word for "word" and since the subjects themselves were often unaware of the underlying concept. The closest Vai equivalent, Scribner and Cole note, literally meant "a piece of speech" or "utterance," and they discovered when they asked an expert in Vai script to divide a piece of writing into such "pieces" what we might expect—that he grouped units together on the basis of the way in which they were used in a specific context. The minimal unit of speech for this expert was not individual words as we know them but clusters of meaning. In studying another West African language, Jack Goody notes that the native term for "word" was best translated as "'bit' of speech" (1977:115), leading him to conclude that "in the beginning was not the word but speech."

Embedded in speech, words can have a symbolic function only to the extent that they are used as part of a symbolic utterance—only, that is, if they are used in a way that removes them from ordinary speech. Yet as the anthropologist Malinowski notes, such symbolic uses of language are hardly the norm: "The manner in which I am using [language] now, in writing these words, the manner in which the author of a book, or a papyrus or a hewn inscription has to use it, is a very far-fetched and derivative function of language. . . . In its primitive uses, language functions as a link in concerted human activity, as a piece of human behaviour. It is a mode of action and not an instrument of reflection" (1923:312). In action, as Malinowski then argues, words are used not to refer to the *concept* of an object but to make the *object itself* appear, to "be handed over to the speaker, or to direct another man to its proper use." In such situations, the meaning of language is to be found, not in symbolic ideas divorced from immediate social context, but in the common interaction that constitutes social life: "A word is used when it can produce an action and not to describe one, still less to translate thoughts. The word therefore has a power of its own, it is a means of bringing things about, it is a handle to acts and objects and not a definition of them" (p. 322).

As a "link in concerted human activity," words often function not as symbols but as icons and indexes with necessary (and not arbitrary) relationships with referents. We use certain words mimetically in certain situations, not because we know or care what they mean, but because we are interested in producing the same results others obtain when they use these words. At such times our use of language reveals the alternate dominance of iconic accommodation and indexical assimilation, as we copy the language use of others in anticipation of being able to produce the same changes in the world. In such situations the central issue tends to be not the truth or falsity of our utterances but, as in presymbolic activity generally, their appropriateness. In this sense, though we readily recognize that our personal names are merely words, we may nonetheless not experience them as symbols; indeed, our names more often act as icons to the extent that we experience them as part of ourselves, just as they are indexes to the extent that we feel summoned by them. This iconic-indexical relationship with words seems to be a universal experience, while relatively few people ever come to regard words as symbols (including most of those who have no contact with written language). We all seem to be born into a world of nominal realism where the bonds between words and their referents seem necessary, not arbitrary. We are all, in other words, akin to the rustic, described by Vygotsky, who could readily accept that astronomers with all their instruments could determine the size and course of the stars. "What baffled him," Vygotsky writes, "was how they found out their names" (1962:129).

Such nominal realism can be considered an elemental form of accommodation. In using the agreed-upon name to refer to objects and people, we are engaged in imitation; the more we are freely aware of the name as an arbitrary convention of our accommodation, the more deliberate and developed our response. Yet if we have little real power over the world, if things seem to happen without our deliberate control and without our making a concerted effort, then we are more likely to sense a causative connection between the word and its referent, a sense that our using the word is a direct means of making the referent appear.

Only the constructive power of assimilation finally allows us to create what is not there and thus to break totally free of accommodative pressures for simple imitation. All of us, however, first learn language during a stage where we have very little assimilative power; this fact as much as any other shapes individual language development. Literacy, for all of us, is initially only potential, affected by two nearly universal aspects of language development—the protracted period of human dependency during the early years of life and the concomitant nurturing offered by the primary caretaker.

An essential part of language growth is the parent's deliberate structuring of context to enhance the possibility of meaningful communication with the child during the first two years of life. Communication between child and caretaker begins in such a supportive environment at birth and continues for many months before the onset of verbal language. As Bruner writes, "Speech makes its ontogenetic progress in highly familiar contexts that have already been well conventionalized by the infant and his mother" (1975:261). Infants make meaningful requests, indicate objects, and even demand that others share experiences with them, all at a prelinguistic level. The mastering of deictic reference, for example, is well established at the sensorimotor level, in the act of pointing, long before its appearance in language.

Thus an infant's first utterance of the word "ball" may be less an act of reference to the class of round objects in the world used primarily for play than merely a new way of requesting or pointing or performing some other activity that has already been mastered at the sensorimotor level. In such an instance, the utterance "Ball!" should really be seen as a holophrase, or one-word sentence, substituting a one-word verbal command that someone hand over the ball for what had previously been communicated entirely by gesture. Likewise, the same holophrase "Ball!" could also substitute for a request for affiliation and confirmation, in which case it might best be translated as the rhetorical question "This is a ball, isn't it?"

Psychologist Elizabeth Bates argues that the missing elements in either holophrase—that is, what is conveyed by either "Give me the X!" or "This is an X, isn't it?"—go unstated by the infant, not just because the infant lacks the syntactic ability to express what exists somewhere within him, but also because what is implied has already been conventionalized at the level of sensorimotor activity and is thus already a meaningful, albeit nonverbal, part of the context. Instead of crying and pointing to a ball, the infant, for example, can now point and utter "Ball!" which is a more effective means of soliciting a desired object when there is more than one object in proximity or when the desired object is not in sight. Such an utterance, according to Bates, is merely another form of Piaget's fifth stage of sensorimotor development, the stage where the infant learns to use a new means to obtain familiar ends. Here the familiar end is the goal of a concerted activity of which language in the form of speech plays only a limited role. The meaning of the language that constitutes the utterance is thus embedded in a specific situation.

In such conventionalized settings with a highly supportive respon-
dent, speech develops, with minimal acts of accommodation (here the
uttering of a single word) able to produce the desired assimilation (here
the presentation of the desired object). We are mistaken, Bates argues,
if we assume that originally language acts symbolically, representing
discrete ideas. "It seems that, initially," she writes, "the child does not
understand or objectify the special vehicle-referent relationship with
which adults link words and things. Rather words are merely a subset
of schemes for operating on objects" (1976:90). In replacing the infant's
cry with an imitation of some sound habitually made by a parent in
association with a desired object, the uttering of "Ball!" operates as an
icon; in making the desired object actually appear, it operates as an
index. In neither case is this use of language symbolic; rather, the
simplicity of both the accommodation and the assimilation guarantees
that the language involved does not transcend the constraints of
context.

Yet this mutual interdependence of accommodation and assimilation
also shows a developmental pattern that is reflected in the growth
both in the individual's freedom from imitation and in the constructive
power of assimilation. Even during the period of single-word utterances,
there is evidence that the infant shows the ability to adjust the utterance to capitalize on shared and thus readily understood information
(Greenfield and Zukow 1978). If, for example, the ball that the infant
desires is already an object of affiliation, so that its identity is not in
dispute, the infant is more apt to utter "Gimme!" instead of "Ball!" to
ensure that an adult responds correctly (I assume that both responses
are available to the infant). The utterance "Gimme!" assumes that the
subject or topic, here the ball, is understood and thus provides the
predicate or comment, while the utterance "Ball!" does just the opposite.
Bates contends that infants will normally direct a holophrase to "that
aspect of a given situation undergoing or subject to greatest change or
uncertainty" (1976:98).

The guiding principle here is the mutual interaction of shared presupposition and new information. During infancy, as well as during
many other stages of life, presuppositions are built directly into language by the manipulation of context. When we ask an infant, "Where
is your mouth?" we are purposely constructing a context that allows
for meaningful communication with only the minimal, nonverbal response of pointing. Between such a deliberately artificial situation and
the world of ordinary speech, there is not necessarily a qualitative
difference. When someone asks us for the date, we regularly respond,
not with a complete sentence ("Today is the———"), but with the
ordinal number; similarly, as Vygotsky notes, someone who first sees
a bus arriving for which a group has been waiting is likely to exclaim

"It's coming." The pronoun "it" is here used, not to take the place of some previously used noun, but to designate information that is shared by the group. The person referring to the "it" that is coming need not even know the common word for bus in the language used by the others who wait with him, just as a child can ask us to throw the "it" that we have in our hands without knowing the common name of that object. Such reasoning suggests that the origins of pronouns may lie in generalized deictic reference and may thus precede those of the nouns that we customarily consider pronouns to be replacing.

Bruner, in "From Communication to Language," draws upon research showing that the average length of children's utterances often increases when they are talking with strangers rather than with their parents, perhaps in compensation for the loss of general presuppositions. Such an explanation may also account for the element of teasing that seems to be so central to language growth, as, for example, when a parent knows that an infant really wants the ball but "misconstrues" the infant's pointing as a request for a book or, after the infant has been forced to utter "Ball," as a request for affiliation. Instead of giving the infant the ball, the parent thus responds, "Yes, this is a ball." Such teasing tactics serve the essential function of teaching the infant the value in being able to overcome a reliance upon shared, unstated presuppositions. "The real accomplishment in pragmatic development," concludes Bates, "will involve learning when NOT to presuppose in order to help a listener who does not have the same assumptions as the speaker. In short, the course of pragmatic development involves learning not to take information for granted and learning to make presuppositions as well as propositions explicit in the surface form of an utterance" (1976:104).

The goal of pragmatic development, in other words, is the same as that of literacy—to be able to communicate with strangers, to be able to break free from the mold of ordinary conversation in order to be able to learn what one would never hear and to assert what one would never say. More than a half century ago, linguist Grace DeLaguna made explicit the connection between general language development and literacy: "The evolution of language is characterized by a progressive freeing of speech from dependence on the perceived conditions under which it is uttered and heard, and from the behavior which accompanies it. The extreme limit of this freedom is reached in language which is written (or printed) and read" ([1927] 1970:107).

The ontogeny of language and the ontogeny of literacy are therefore part of one and the same process. We must gain the ability to make increasingly specific our presuppositions, that which we share with those around us, so that we can communicate with new audiences. We must learn not just to comment upon some topic of mutual interest but to *create* that topic in language for people for whom it does not exist. The topic, in other words, must ultimately come to exist in the discourse itself rather than in the context; only then can we speak of the symbolic nature of discourse, for only then is the meaning free of the synecdochic, part-whole ties to preexisting meanings embedded in our social relations. Thus the word is a symbol, rather than an icon or an index, not because of some inherently arbitrary relation to its referent, but only because it is part of an utterance whose essential purpose is neither iconic or indexical, that is, only because it is part of an utterance that seeks to identify a new understanding rather than merely to remind others of something already known. The word is a symbol only when it is part of a symbolic utterance.

The notion of words-as-symbols can thus be seen best as coming at the end rather than at the beginning of a long developmental process, either for individuals or for society at large. Accordingly, Cassirer is wrong in seeing signs generally as standing in fixed opposition to the "actual flow" of consciousness, since the pattern of accommodative-assimilative response, of which the word-as-symbol represents the apex, begins with our earliest efforts to adapt the world to ourselves and ourselves to it. These patterns of adjustment necessary to maintain life, what Piaget calls "schemata," constitute the precursors of all representational activity, for the iconic image of any accommodation is inevitably linked with the indexical effect or payoff of any subsequent assimilation.

Language use in infancy, although constituted of words, lacks an awareness that such signs arbitrarily denote a world of potential, rather than necessary, meaning. Signs attain the "definite ideal *meaning*," which Cassirer rightly sees as crucial to human development, only when, through the process of increasingly complex accommodations and assimilations, they come to represent not what is but only what is possible. The word is thus the fulfillment of the search for a theoretical understanding of the world, an understanding that as Vygotsky notes begins not with language itself but with human activity:

> To the Biblical "In the beginning was the Word," Goethe makes Faust reply, "In the beginning was the deed." The intent here is to detract from the value of the word, but we can accept this version if we emphasize it differently: In the *beginning* was the deed. The word was not the beginning—action was there first; it is the end of development, crowning the deed. [1962:153]

Vygotsky does not clarify the point that deeds and words coexist and that our skill at using either one develops simultaneously—we act in the world both by doing and by saying things, just as both our actions and our words can transcend the boundaries of the present and can achieve symbolic standing. We can create symbolic meaning with our deeds just as we can act with our words. The final goal, however, for both our words and deeds is the same—the construction of that which transcends the limits of the present. Language does have a special role to play in this process; before considering this role, however, it is necessary to establish more specifically the relationship between the potential symbolic nature of words and the attainment of theoretical knowledge generally, for if words are a beginning in any sense at all, it can only be a beginning to distinctively human mental operations.

III. In the article "The Contexts of Comprehension," psychologists David Olson and Nancy Nickerson stress something very important about the difference in how children and adults process language. Adults can quickly see that the two sentences "John hit Mary" and "John was hit by Mary" are not equivalent, indeed that they oppose each other even at the level of syntax. That is, even if some nonsense word was substituted for "hit," so that it was impossible to know just what John did to Mary, most adults could readily recognize the two sentences as describing opposite conditions. Children, on the other hand, have a more difficult time drawing this conclusion and, unlike adults, have a relatively easier time recognizing the equivalence in the two sentences "John hit Mary" and "Mary was hit by John." Olson and Nickerson explain the difference in language processing by hypothesizing that children actually go to the trouble of picturing the state of affairs described in each sentence and then comparing these pictures to see if they are the same or different. While adults can use the propositions as a basis for drawing a logical inference, children feel a greater sense of necessity between the wording and the state of affairs the proposition posits.

Similarly, as Olson notes elsewhere, when asked "Which is not bigger?" children tend to respond "X is not bigger" rather than "Y is smaller," and when asked to verify that "7 is not an even number," they do not transpose the proposition to mean that "7 is an odd number" before responding. Other research described by Olson explores this developmental component in modes of comprehension by asking children to respond to questions based on the following sentence: "The elephant is ahead of the giraffe and the camel is behind the giraffe." Six-year-olds generally had great difficulty in drawing any conclusions until the researcher embedded the same information in a richer, more meaningful context.

The elephant, the children were told, was ahead because the "giraffe's long neck kept getting tangled in the thick branches," and the camel is behind the giraffe because he "frequently stopped to eat." With this new information, the six-year-olds could draw the correct inference even though the reasoning demands remained the same. First graders, it seems, are indeed capable of deductive reasoning but only when the propositions are embedded in sufficiently concrete, albeit fictitious, circumstances. Children seemingly fail to grasp that such propositions have meanings independent of the contextual restraints represented by their specific wording.

At stake here is the cognitive difference between an inference based upon logical analysis ("If $A<B$ and $B<C$, then $A<C$") and one based upon a comparison of specific items in a concrete situation ("If my cousin is taller than my uncle and if my uncle is taller than my aunt, then my cousin is also taller than my aunt"). The logical operation has a universality of application that the specific operation lacks, although there are undoubtedly occasions when it is preferable to pay close attention to a specific situation and to ignore what seems logically necessary. The real superiority of the logical operation to the specific comparison lies not in some abstract hierarchy of cognition that always places the general over the concrete but only in the matter of the relative degree of freedom that exists within the individual between the sign or verbal proposition and what it signifies.

The limitation of the six-year-olds' response is manifest in the relative predominance of accommodation, evident in their reliance upon a specific context to shape meaning. The first graders' dependence on context finally prohibited them from responding correctly to the un-adorned syllogism about the camel, the giraffe, and the elephant; they needed a richer narrative to compensate for their relative inability to constructively transform whatever was given. To six-year-olds, in other words, these propositions function primarily as indexes in positing what seems to them to be a necessary state of affairs.

These children nevertheless do draw the correct inferences, thereby revealing an important distinction between logical thought per se, a product of the synchronic analysis of human behavior, and symbolic representation, a product of diachronic analysis. In commenting upon the difficulty or, more accurately, the refusal of the peasants in Luria's study to base their response upon inferences drawn from hypothetical conditions, anthropologist Edwin Hutchins (1980) notes that the peasants are not acting illogically; on the contrary, he argues, they are adhering to a different and even higher order of logic. They believe that, despite the information contained in the syllogism, one can only know the color of a bear by seeing it and furthermore that one can see a bear only by being in its area of habitation. Thus, using a form of reasoning known as *modus tollendo tollens* (reasoning from the consequent), they logically conclude that they have no way of knowing the bear's color. To respond to the syllogism directly requires a less demanding form of reasoning, *modus ponens* (reasoning from the antecedent).

Such abstract ranking of levels of reasoning, however, ignores the issue of the historical development of symbolic representation, instead postulating an abstract hierarchy of types of logical reasoning. The issue here is not whether one action is always more complex or more abstract or somehow better than some other action but only whether within a specific historical context one action represents the progressive efforts of an individual or group to free themselves from the constraints of the given by making their accommodations more deliberate and their assimilations more constructive.

To see this ontogenetic pattern in symbolic representation, it is helpful to draw upon the vocabulary of classification used by both Vygotsky and Piaget, a vocabulary that has important parallels with Peirce's levels of signification, discussed above, and is thus particularly helpful in revealing the unified ontogenetic roots of symbolic language, thought, and action. The three levels of classification that Vygotsky defines are "heaps" or "congeries," "complexes," and "concepts." The items in a congeries are selected on an ad hoc basis without the consistent application of a single rule. The congeries functions as an icon in being little more than a random sampling or miniature of what is given, and as such it represents the least deliberative accommodation and least constructive assimilation—the individual merely copies what is there.

With a complex, however, there is a more deliberative and more constructive response, as the individual reshapes what is given accord-ing to a rule drawn from everyday experience. Objects in complexes, writes Vygotsky, are grouped "on the basis of their participation in the same practical operation—of their functional cooperation" (1962:63). As Luria discovered in the peasant's insistence that a log must be grouped with three common tools, a psychological tension binds ob-jects in a complex, making them highly resistant to alteration. The "functional cooperation" between objects, their actual association in ordinary experience, gives complexes the causal connective power of indexes.

The formation of the concept involves the most freely chosen accom-modation and the most constructive assimilation, for the individual is grouping objects according to a rule that reflects, not actual practice, but a logical possibility. The concept thus corresponds to the symbol in being grounded upon what can be rather than what is. The concept, as such, is the cognitive equivalent of literate discourse, just as the complex is the equivalent of synecdochic communication. "A complex,"

Vygotsky explains, "does not rise above its elements as does a concept; it merges with concrete objects that compose it" (1962:65).

Of crucial importance here is the role of classifications that fall between the complex and the concept. These classifications appear to be concepts because they are selected on the basis of the application of a single rule, but they never really break the bond with ordinary experience because the rule itself is grounded in experience. Called "pseudoconcepts" by Vygotsky and "figural collections" by Piaget, these classifications are routinely formed by seven- and eight-year-olds when they group blocks on the basis of some overwhelming similarity such as color. Such children may seem to be performing symbolically; they lack any sense, however, that any particular grouping represents only one possible subset of the original. Pseudoconcepts are not as freely chosen and, as such, are not indicative of a fully developed mode of accommodation. On the surface, pseudoconcepts do resemble concepts; they are connected with indexes only by the persistence of a psychological bond with ordinary experience. Instead the classification is imposed upon the individual by circumstances that draw attention to some outstanding, dominant characteristic. Pseudoconcepts are thus to be distinguished from concepts in the same fashion that symbolic language is to be distinguished from synecdochic language and praxis from practice—that is, only in terms of a diachronic model of human development.

It is important to recognize that pseudoconcepts can sustain cognition without full development of literacy. We can maintain a high level of performance both in actions and in speech at presymbolic level. As Olson's research reveals, children can perform the same logical functions as adults. Children do so, however, in a significantly different fashion—by working with the representations of the texts as if they posited real, rather than hypothetical, situations. As Bruner and Greenfield note, the very act of classification changes its character when an individual describes a grouping in a complete sentence rather than by a single descriptive term: "The embedding of a label in a sentence structure indicates that it is less tied to its situational context and more to its linguistic context (1973:47)." As Olson and Nickerson conclude, this new competence in negotiating meanings divorced from context develops fully only with literacy:

Our tentative conclusion is that the form of human competence involved in constructing a practical model of reality, in making predictions intelligently on the basis of that model, and in assimilating oral statements to that picture of reality, is the general and largely universal possession of mankind, young and old, literate and illiterate. But the form of human competence involved in the ability to confine interpre-

tation to the information explicitly stated in the text and to then oper-
ate on that meaning to derive logical entailments is tied largely to the
development of literacy. [1978:165–66]

Thinking with pseudoconcepts, that is, with images that bear an indexical connection to ordinary experience, is the norm, to be surpassed
in most cases only with practice creating and comprehending verbal
meanings in the theoretical world of texts.

Verbal meanings embedded in writing are in all ways less constrained
by indexical involvement with specific patterns of social experience
and thus represent one of the most readily available and least threatening forms of symbolic activity. While practice with reading and
writing does seem to facilitate the formation of concepts, the connection here, as we say in relation to collective cultural development (for
example the case of classical Greece considered in chapter 3), is inter-
dependent rather than causal. Indeed, the relationship between development, either individual or collective, and literacy seems to be much
more complex than is normally believed. Havelock notes, for example,
that "all the essential features of the Greek way of life" (1976:5)—
which many have attempted to explain as the result of literacy—seem
to have been in place some three hundred years before the development
of the phonetic alphabet. It thus seems possible to seek a unified
explanation for the growth in both cognition and literacy in the deepening of a more general capacity to form symbolic representations. In
this sense, the historical formation of phoneticism in ancient Greece,
identified by Goody and Watt as a cause of cultural development, must
itself be explicable in terms of a collective increase in the Greek power
of symbolic representation.

Such an approach entails recognizing the essentially iconic nature of
most prephonetic forms of transcription. While pictographs, for example, are clearly icons, it is possible to argue that even syllabaries are
representations of naturally occurring sound patterns in any language.
Syllabaries are therefore like complexes in representing what does exist
rather than what might. Phonetic systems, however, deal with the
possible rather than the real by dividing naturally occurring sounds
into component parts—vowels and consonants—that exist as abstrac-
tions or theoretical possibilities. Vowels are not sounds but only instructions for shaping one's mouth in the formation of sounds, "at-
omized phonemes," in Havelock's phrase, while consonants are really
nothing more than instructions for manipulating the column of air by
means of the vocal cords, tongue, palate, lips, and nose.

By reducing all natural sounds to these two abstract components,
phonetic systems are able to reproduce in written form all the words

of a given language with a minimum number of signs. The Greek invention of the phonetic alphabet, in Havelock's words, "consisted in pursuing the analysis of linguistic sound to the level of complete abstraction" (1971:10). The development of phoneticism itself, in other words, is already an immense cultural achievement, one as suitably considered an effect as a cause. It is thus quite possible that literacy had radical consequences in Greece (and not in other places) for at least partly the same reason that Greece was the first area to develop a phonetic alphabet. The same confluence of factors, most likely related to a deepening of praxis generally, may well have been responsible for the development of symbolic representation that led first to the emergence of phoneticism and then to the highly symbolic uses for written language analyzed by Goody and Watt as the "consequences" of literacy.

Identification of the specific factors that made Greece develop differently would take us far beyond current concerns, but we can nevertheless mention the general constellation of factors that is likely at work whenever literacy flourishes. Perhaps the central ingredient is a pattern of historical change that leads to people's obtaining an increasing sense of both their power to change the world and the difficulty of that task; a deepening of the process of interaction between human will and natural resistance seems to be most necessary—a process that is apt to be reflected at the collective level by an emerging sense of human progress. For literacy to develop, action must lead to changes in the world and, in turn, to changes in our thoughts and our words; eventually, in all three modes of experience—in activity, thought, and language—we must form increasingly deeper symbolic representations, reflecting our greater power to change both the world and ourselves.

It is perhaps possible to attain the level of symbolic activity purely in action (that is, as praxis) or purely in symbolic, nonverbal thought, but such achievements are apt to have only local significance, if even that. Only the achievement of language gives wide accessibility to such attainments, placing them within the reach of all who hear our voice, and only the accomplishment of literacy gives them real universality. As a result, only the accomplishment of literacy fully humanizes the transformative, liberating power of symbolic experience by allowing all of us to build new worlds from the very words, and hence the very experiences, that bind us to this one. Only in the necessarily conditional mood of literacy do we deal with the most fully realized and the most accessible world of possibility.

IV.   Although speech, thought, and practical activity all have distinct origins, they converge and intermingle very early in life in such a way as to produce a distinctive individual and collective human experience with only minimal parallels in animal life. While the other higher primates are also capable of problem solving and even of certain kinds of language use, the constructive, purposeful interaction of thought, action, and language, even within the first eighteen months of human life, transforms quantitative differences between human and animal life into qualitative ones. Unlike apes, Vygotsky notes in comparing cognitive development in children and higher primates, "children not only *act* in attempting to achieve a goal but also *speak*" (1978:25). Speaking is an inestimable aid for children, especially in problem solving, since their constructive assimilative power to change the world in language is generally far greater than their comparable power to change the world in action. Speaking thus aids the child in the formation of possibilities. Alternatives can arise with great ease, and these alternatives, once they have been transformed from unexpressed or felt thoughts into verbal form, also gain something of immense consequence—a public character. Our constructive verbal assimilations, no matter how fanciful or how idiosyncratic, nevertheless remain indirectly grounded in the common social experience of speech.

In *Mind, Self, and Society*, the American social psychologist and philosopher George H. Mead locates the basis of all meaning and in turn the basis of distinctive human mental life in our ability to indicate to ourselves what we can indicate to others. The status of a stick, for example, changes radically when, instead of using it to stun an animal, we use it to point so that another person can see the animal, for only then does the stick act as a sign facilitating an act of communication, indicating to the other person what it also indicates to us. Likewise, the speech of a child that accompanies problem-solving efforts creates a meaning that the child can generate and to which it can respond. Such meanings, it is important to note, are inherently public, belonging exclusively to neither party but to the world of shared social experience; precisely this public character of signification explains the great advantage of language use and eventually of literacy over other forms of praxis.

As Vygotsky writes, "Thus, with the help of speech, children, unlike apes, acquire the capacity to be both the subjects and the objects of their own behavior" (1978:26); that is, the language of children during problem solving is not just another form of behavior but also an early manifestation of the species-specific activity of the symbolic language use that constitutes literacy. The children's words create possibilities that are meaningful and hence independent of either their immediate intention or their will; children are capable of responding to this verbal meaning as if it were something outside themselves, as if, in other words, it were only a hypothetical possibility. The individual's use of public language to facilitate a private dialogue constitutes what Vygotsky identifies as the characteristically human activity of *inner speech:*

The greatest change in children's capacity to use language as a problem-solving tool takes place . . . when socialized speech (which has previously been used to address an adult) *is turned inward.* Instead of appealing to an adult, children appeal to themselves. . . . When children develop a method of behavior for guiding themselves that had previously been used in relation to another person, when they organize their own activities according to a social form of behavior, they succeed in applying a social attitude to themselves. The history of the process of *the internalization of social speech* is also the history of the socialization of children's practical intellect. [1978:27]

The pattern of development that Vygotsky identifies here stipulates that individual mental constructs exist first at the social level. Our distinctive mental life, then, is largely the result of our applying this "social attitude" to ourselves. As Mead explains, our sense of self-identity is the result of our being able to incorporate within ourselves this generalized sense of the community. "What goes to make up the organized self is the organization of the attitudes which are common to the group" ([1934] 1967:162). The values, attitudes, and, most important, the language of the group must exist within us; our ability "to call out" in ourselves the responses of the group Mead defines as "self-consciousness" (p. 163)—that is, our sense of the extent and the limit of what we share with a group and hence of what we do not share as well.

Self-identity, as defined by Mead, is not to be confused with individual differences. We are all different in the sense that all natural phenomena exhibit individual difference; Mead here refers not to these differences in themselves but to our awareness of them. For Mead, we experience ourselves as individuals only as part of an internal dialogue between ourselves as subjects, what Mead identifies as "I," and ourselves as objects, what Mead identifies as "me," or the direct awareness of ourselves that we can attain only by adopting the attitude of others. We are thus both the "me" that we can know only by adopting the attitude of others and the "I" that acts in response to our internalized sense of others. The "I" ultimately solves the problem and ultimately speaks out, inwardly or in public, giving public form to private experience. The "I" acts by incorporating potential courses of action within our sense of what is appropriate and what is possible in our social world. The "I" reacts to the sense of the other that we incorporate into ourselves as "me" and that in turn is subject to constant change in the course of our internal dialogue with ourselves. Without this internal dialogue and thus without a language that provides some regular means of communication with others, we would have little sense of our ability to respond to others in a systematic way and thus little sense of how we are both like and unlike others. Without the interactive communication with others provided by language, in other words, we would lack a full sense of ourselves as objects, a full sense of ourselves as "me."

Language is capable of generating alternative responses and in so doing of incorporating a new world of possibility within our generalized sense of the other. In inner speech, the "me" acts only at the level of possibility, reformulating and expanding the boundaries of what is given, establishing the context against which the "I" can continue to act. While language thus ensures the existence of "me," the dialogue initiated by the "I" plays the critical role in ensuring the growth and development of our distinctive sense of self-identity. The nature and extent of the dialogue we carry on with ourselves is therefore crucial for the development of self-identity; since we carry on this dialogue only by adopting the responses of others, the nature and the extent of the dialogue and, by extension, the likelihood that the sense of self-identity will develop in large measure reflect the attitude and practices of the community.

With little power for constructive assimilation, however, a group will tend to possess a level of integration that both effortlessly trans-poses the world (often through magic and ritual) and quietly imposes strict patterns of imitative practice. Such a level of cultural integration, when incorporated within an individual, does not promote the internal dialogue between the alternatives of "I" and the responses of the "me." Since assimilation occurs with such ease, the alternative forms of being generated by thought and language take on the status more of ominous threat than of theoretical possibility. As J. C. Carothers has observed, even the recognition of the purely *linguistic* nature of speech requires a certain level of cultural organization: "In those societies—apparently comprising most, if not all, nonliterate societies—in which verbal thoughts are seen as having power in the real world in their own right, no clear distinction can be made between thought and action, words are regarded as being of the same order of reality as the matters and events to which they refer, and thought is seen as being 'behavioral' in the same sense as any other type of action" (1959:312).

Even the physical source of the dialogue is in dispute, as the pressures for accommodation leave little room for anomalous thought (or what Carothers suggests is really experienced as anomalous "action") within the individual. There is no awareness of the boundary of the generalized other and hence no sense of where the other ends and where our sense of our self, the distinctive "me," begins. Instead there are only two, seemingly external, forces at work—the legitimate force of the community and the seemingly demonic voices of anomaly. In the view of anthropologist Godfrey Lienhardt, what is lacking is "our popular modern conception of the 'mind' as mediating and, as it were, storing up the experience of the self." Such is the case, he writes, for the Dinka of the Sudan:

There is for them no such interior entity to appear, on reflection, to stand between the experiencing self at any given moment and what is or has been an exterior influence upon the self. So it seems that what we should call in some cases the "memories" of experiences, and regard therefore as in some way intrinsic and interior to the remembering person and modified in their effort upon him by that interiority, appear to the Dinka as exteriorly acting upon him, as were the sources from which they were derived. Hence it would be impossible to suggest to Dinka that a powerful dream was "only" a dream, and might for that reason be dismissed as relatively unimportant in the light of day, or that a state of possession was grounded "merely" in the psychology of the person possessed. They do not make the kind of distinction between the psyche and the world which would make such interpretations significant for them. [1961:149]

The Dinka, like any group, no doubt consist of a great many distinctive individuals, and their language, like all human languages, allows them to internalize the attitude of the group as the basis for self-identity. Yet according to Lienhardt's account, any sense of the theoretical nature of the verbal actions of the "I" seems to be lacking—any sense, in other words, that these verbal actions do not really constitute a constructive assimilation of the world. The speech of the Dinka's "I" therefore represents a very low level of symbolic representation, indeed exists more as index than as symbol, in carrying with it a constant sense of causal connection with the world it signifies, as if the utterance of certain words or even the thinking of certain thoughts is itself enough to precipitate the magical transformation of the world.

Ultimately the level of symbolic representation in the speech of the "I" matters as much as the fact that the "I" speaks at all, for only in the increasing depth of representation of language (and thought and action) does one experience the full range of possibilities for and constraints on human action—and only in our awareness of such possibilities and such constraints can we experience a sense of individual autonomy. The speech of the "I" must not be seen as changing the world directly (or if it does, it will be seen as threatening and will likely not even be recognized as our own); instead it must be seen as positing only possibilities for accommodation.

Gay and Cole report that, among the Kpelle of Liberia, court cases are decided by the party who can "produce an argument demonstrating conformity to tradition that the other party cannot answer" (1967:24–25). It is not necessary for this argument to be supported by evidence or for it to display some internal plan consistent with rules of formal logic. There is no sense of the legitimacy of what educator Lawrence Kohlberg calls "postconventional morality" or of the belief in "autonomous moral principles that have validity and application apart from the authority of the groups of persons that hold them" (1970:110). There is no sense, in other words, that a body of knowledge or a world of texts embodies symbolic meanings not embedded in the collective practices of the group.

Such a notion of *truth not founded on experience*, according to anthropologist Rodney Needham (1972), forms the basis of the Pauline transformation of the Old Testament sense of "faith" into the New Testament sense of "belief." Old Testament faith, Needham contends, called for loyalty and obedience to God because of what He has done for his people in the past. New Testament belief reverses the process, making the very acceptance of what God does (specifically the miracle of the Resurrection) an act of faith independent of ordinary experience. The act of faith at the center of Christianity reflects the literate paradigm of forgoing synecdochic truth for symbolic truth. Nor is the connection between literacy and Christianity merely coincidental. As Jack Goody notes, the spread of literacy generally leads to the emergence of the notion of belief as faith in precisely that which is not supported by practice. The religions that preach conversion and exclusion, Goody writes, are "all religions of the book" (1968a:2).

Only literacy ensures the existence of complex and developed systems of belief apart from the rule of the group. As Zygmunt Bauman observes, only the objectification of such knowledge as part of the search for truth offers us recourse from the pressures of consensus. Belief as we know it today is only possible where truth can exist apart from consensus and apart from the world of ordinary discourse upon which consensus is built. "The pursuit of true understanding as distinct from ordinary agreement," writes Bauman, "must therefore detach itself from everyday discourse and seek its own rules elsewhere" (1978:232). There is no more likely "elsewhere" than in the reading and writing of texts, that is, in literacy itself. Only in the development of literacy is there full opportunity for the development of the "I," for only in such activities are our assimilations as fully "constructive" and our accommodations as fully "deliberate" as is possible. The "I" can build symbolic representations without recourse to language, just as it can speak without recourse to symbolic representation, but in neither case does it attain its full depth of purpose—in the former case, there is no dialogue and no sense of an emerging "me," and in the latter case, there is dialogue but no sense of a distinctive "I."

Literacy guarantees precisely this sense of a distinctive "I"—that is, not the distinctive actions of an individual but an individual's sense of his or her own distinction. We are ordinarily aware only of the "me," the generalized other within us. As Mead notes, the "I" always escapes our notice—no matter how fast we turn around, we cannot catch a glimpse of ourselves acting. We cannot be the objects of our creative selves, except retrospectively. According to the philosopher R. G. Collingwood, "All knowledge of mind is historical" (1956:219); we can only know the "I" that is ourselves as a historical figure—a figure not unlike the authors whom we study closely for years and whom we can finally come to know intimately but indirectly through our study of their efforts to overcome the limits of synecdochic experience. As readers we appropriate the world of others, widening our sense of both the possible and the real. Our "I" joins with the "I" of others, creating worlds of possibility that can be grasped only historically by an emerging "me." As writers we act, creating what is as yet not part of our social self. All literate experience is exploratory, directed to that emerging self, offering it a world of new possibilities, including the most valued possibility of all, that of self-knowledge. Nowhere else are the struggles of the self as ever present and as elusive as in our reading and writing of texts.

An incident in the Truffaut film *Small Change* reveals a great deal about the role language plays in human development. A teacher listening to a young student make his way haltingly through a recitation from Molière is unexpectedly called from the room. Crossing the open courtyard below, he hears the same student reciting the passage again—only this time effortlessly, in the dramatic, exclamatory style of the teacher himself. Freed from the ordinary social restraints and thus able to transform the world according to his own desire, the student chooses to imitate the teacher and in so doing ironically gives life to his own expressive powers. And so it is when we read and write—we free ourselves from the constraints of speech, becoming what we are not so that we may ultimately become what we are. Of all human activities, literacy is perhaps best able to fulfill the self, for in acts of reading and writing, as nowhere else, we are free to become, and so in a sense, to call into being, what is not. In acts of literacy, more than in any other activity, we are free to take full possession of what George Steiner calls our greatest tools—"the uses of language for 'alternity,' for mis-construction, for illusion and play." With this stick, Steiner writes, we have "reached out of the cage of instinct to touch the boundaries of the universe and of time" (1975:224).

Part Three

Progress

5

The Revisionist
Critique

I. Even though reading and writing occupy places of comparable importance in most models of literacy, this conceptual equality disguises a very real disparity in the actual attention and resources generally devoted to instruction and research in reading and writing. Differences in the level of funding for support of research, for example, have favored reading over writing by as much as a hundred to one (Graves 1978:12). The phrase "literacy education" in the United States today actually refers largely to the efforts of educators, often based in schools of education and acting with government sponsorship, to achieve minimal levels of *reading* skills throughout the population, with special emphasis given to those people no longer attending school.

A work like Carroll and Chall's *Toward a Literate Society*, for example, which at first glance appears to be a general discussion of literacy, on closer inspection proves to be the report of a federally funded project conducted by a national reading organization concerned largely with the problems of adult illiterates. In the introduction to this work, Carroll and Chall describe the primary "literacy" task confronting our nation as "ensuring that every child arriving at adulthood will be able to read and understand the whole spectrum of printed material that one is likely to encounter in daily life" (1975:8). The absence of any mention of writing merely continues a long-standing tradition.

In recent years, the National Council of Teachers of English (NCTE), aware of this imbalance both in general concern and, perhaps more important, in overall funding, has made a deliberate effort to widen our sense of literacy education. In part the NCTE was concerned to have Congress define writing as a "basic skill," along with reading, in Title II of the Elementary and Secondary School Act of 1965. The term "basic skill," however, is an ideological crux, meaning different things to different people. As noted in chapter 2, the NCTE has long been a supporter of the problematic model of literacy and thus has its own particular sense of what makes writing "basic."

In outlining the scope of a basic skills writing program, for example, 115 the NCTE advocates telling students to write in "many forms" (1979) and then lists nine examples—seven (poems, stories, scripts, journals, essays, notes, and summaries) having for most students little practical application outside of school or, in some cases, even outside the English classroom, and the other two (letters and reports) often playing a less important role in students' lives than something as common as completing a job application form. In fact, filling out such an application, a prominent feature on life-skills proficiency tests of basic skills, is conspicuously absent (perhaps intentionally so) from the NCTE guidelines for a basic writing program. In calling writing "basic," the NCTE is likely indicating that writing is somehow essential for being educated, whereas the original drafters of the legislation may well have meant something quite different—namely, that "basic" skills referred only to what one needs to know in order to gain employment and to function successfully outside school.

Indeed, writing has quite probably fared poorly compared with reading precisely because its practical or "basic" value outside school is suspect. The nine forms of writing listed by the NCTE may speak eloquently to the developmental needs of all students, but they speak to the practical needs of only a minority. Writing is undoubtedly of considerable practical value to this minority—many professionals do prosper because of their writing skills—but it is very difficult to establish just how many or to determine whether other communication and social skills involving the ability to interact with coworkers or superiors may not be even more important.

While the recent growth in university-based technical writing programs does reflect the need for a certain level of writing competence in technically trained individuals about to enter the workplace, the very need for such programs calls into question the level of writing skills of students who have completed all the general requirements in writing that have been incorporated into the regular high school and college curriculum. Do we need advanced writing programs like those in technical writing because the literacy demands of the workplace are today so much greater than the ordinary requirements of a college writing program, or do we need them because a growing number of technical and professional positions now really do require effective writing skills? Could the proliferation of technical writing programs possibly reflect not so much the general importance of writing in the workplace as the cost effectiveness of using writing specialists, thereby relieving engineers and scientists from having to be able to express themselves in clear, ordinary prose?

The fact remains that, while only a minority of workers will ever create literate texts, many more will have to master minimal coding skills. Such coding skills, however, often entail little more than the mastery of the most elemental record-keeping devices, for example the ability to complete checklists, timecards, inventories, and invoices—tasks that can often be completed with tallying skills that fall considerably below even the minimal requirements in penmanship, punctuation, and spelling necessary to "write" according to the unproblematic model of literacy.

Without longitudinal studies of writing in the workplace comparable to the work on reading done by Thomas Sticht and the Human Resources Organization, one can only speculate about the vast numbers of people, some perhaps even in positions of considerable authority, who are generally meeting the demands of their work with little ability to organize their thoughts and experiences into coherent texts. After all, one person hired as a staff writer can write the instructions that millions of people will have to read—and as Sticht confirms, many people can still do their jobs satisfactorily, even if they cannot read the instructions intended for them, much less write such instructions themselves. It is therefore difficult to take issue with the assertion of language educator Walter Loban that "most men and women functioning effectively in their communities use writing much less than reading." Such an approach to literacy leads Loban to a seemingly logical conclusion: "Writing cannot be justified on any utilitarian basis and inasmuch as successful instruction in writing is a heavy burden for both teacher and learners, we need to reconsider its place in society for those who do not lead lives in professions or activities where writing is necessary" (1978:98).

Loban's words are an attack, not on literacy per se, but on our tendency to overvalue it, to believe that teaching people how to write and (although this point is not stated) also how to read will somehow almost magically transform their lives. There has traditionally been little doubt among educators that, as noted in the UNESCO statement of 1976, "the very process of learning to read and write should be made an opportunity for acquiring information that can be immediately used to improve living standards" (1976:10). Here we have the basic assumption at the heart of much of the ideological controversy surrounding literacy—the belief that a people's achievement of a certain facility with written language relates closely to their efforts to improve the quality of their lives. As historian Harvey Graff writes, "The rise of literacy and its dissemination to the popular classes is associated with the triumph of light over darkness, of liberalism, democracy, and of unbridled progress" (1979:xv). Graff here is writing revisionist history, attacking what he labels the "literacy myth," the naive belief that literacy promotes social mobility. Using extensive documentary evidence from nineteenth-century Ontario, Graff demonstrates in convincing detail how the widespread increase in literacy among the working classes did not result in significant alterations of the prevailing social structure.

The problem with Graff's work has nothing to do with the accuracy of his data or conclusions or even with the question of the complex historical relationship between class and literacy (a subject to be considered in chapter 6) but only with Graff's insistence that his study is mainly about "literacy." For Graff as for many other historians, a person is "literate" who can perform some minimal coding tasks (like signing his name) or has demonstrably attended school a limited number of years. Working with such a definition, Graff manages to prove only that the growth in education in the nineteenth century that led to the achievement by many more people of coding skills sufficient to make them no longer "illiterate" did not result in significantly greater social mobility. The real myth in Graff's work, therefore, is not that "literacy" produces mobility but that working-class children really emerged from nineteenth-century schools as *literate* or, given the limited scope of nineteenth-century schooling, that *literacy* was ever really a goal of nineteenth-century public education in Ontario any more than it was anywhere else. It is, quite simply, a myth to suggest that most students emerged from their few years of schooling as proficient creators and comprehenders of texts.

The basic issue in Graff's work is not literacy but schooling and its essentially conservative nature in reproducing the social order. Is it possible, we might ask, that one reason schooling failed to effect greater social change was that it was unable or unwilling to make all students fully and equally literate? This question, however, Graff is not interested in pursuing. Rather, Graff gains an ideological advantage broadly in calling his assault on the conservative nature of schooling an attack on literacy. To distinguish schooling and literacy, as the problematic model requires, is to raise the possibility both that schools are often very conservative precisely because they fail to promote literacy and furthermore that the ideal of literacy, unlike the actual practice of much reading and writing instruction or schooling generally, is potentially liberating.

The adoption of such a position might entail calling for the reform of current education practices in order to ensure that more students would become truly literate. Yet such a conclusion is precisely what Graff, working from a revisionist perspective, wants to avoid. Instead he wants to include literacy in his attack on our current understanding of the past, for ultimately his target as a revisionist historian is not just nineteenth-century schooling (far too easy a mark) but our misplaced faith in literacy itself and the overly narrow Western model of development that it seeks to promote. Graff, like many other critics of literacy, is interested in debunking literacy (always defined according to the unproblematic model) as a mechanism for attaining either economic development or, more to the point, a higher form of cultural life.

Goody and Watt in their 1963 essay tied literacy to the emergence of skepticism and the general drive for objectivity of thought that destroyed the emotive, collective nature of oral life. They argued that Western skepticism—the divorce of theology from religion, philosophy from faith, history from myth—was the necessary consequence of literacy. For Graff and for other contemporary critics of literacy such as Scribner and Cole, such thinking is a form of reification and mistakenly equates the mastery of coding skills, a limited accomplishment, with both individual and collective development. Writing is then seen as a fetish capable of magically transforming the world, not for what it really is, a valuable but limited means of encoding language. Underlying this belief, in the eyes of such critics, is an ethnocentricity that tends to judge everything foreign according to narrow, culturally biased standards: other people and other ways of life are less developed than our own, hence inferior, and thus our political and economic exploitation of them is justified. The critique of literacy has as its goal the unmasking of our belief in the transformative value of literacy; it seeks to expose this belief as an apology for the Western domination of both specific ethnic minorities and indigenous practice generally.

This critique, however, is not without its own mystifications and contradictions. While the attack upon literacy is part of a larger critique of Western culture, it is also an important part of that literate tradition, perhaps even the most important part of that tradition. In attempting to formulate in writing verbal meanings that extend our current understanding of the world, the critique of literacy is itself inevitably a significant and invaluable, albeit ironic, act of literacy. Indeed, for the last hundred years, while literacy was becoming increasingly important in schools and society at large, many of the great innovators in various fields of cultural study, notably linguistics, art history, and anthropology, were writing books that enlarge our own understanding of human culture by revealing the hidden complexities and beauties in the verbal and general communicative practices of those who cannot read and write. To grasp truth, our greatest scholars of the last two centuries have been telling us, we must transcend the limits of our present understanding—truth ultimately exists *over there*, somewhere beyond the here and now. Such perpetual revisionism is part of the very nature of literacy; it is a kind of Hegelian dialectic where the present is always being sacrificed for an image of some other, emerging sense of truth.

Thus it is hardly surprising that the most literate studies of language during this period have been largely concerned with analyzing, not what was self-evident in the texts themselves (that literacy gave one a certain power over the world), but what such a self-evident belief denied—namely, that nonliterate communication had its own special value. Just as one of the first uses of writing in ancient Greece was to study in rhetoric certain formal aspects of speech, so one of the major uses of literacy during the past hundred years has been to study speech and the vast store of verbal material that has been preserved in oral form. The ideological attack on literacy begins with the rejection of the tyranny of the written word and the recognition, initially on the part of linguists and cultural anthropologists and, more recently, of language educators generally, that speech has its own complexity, as great as, if not greater than, that of writing itself.

II. For most of the nineteenth century, linguistic research focused largely on the study of written language. It took the form of philology, a branch of study that has the same origins as hermeneutics and, at least initially, had the same concern with historical understanding. During the early decades of the century, philology, especially in Germany, was transformed by the recognition that documents from the past, if studied in a true historical spirit, would bring in greater knowledge of distinctly different human experiences. Philology thus developed as a means of analyzing the language of texts as part of the larger process of understanding distant cultures; written texts were valuable as expressions of the past that could be resurrected in the hermeneutic act of reading.

By the end of the nineteenth century, however, philology had lost much of its original idealistic inspiration, becoming less of a search for a hermeneutic understanding of the past (hidden in texts) and more of a limited, scientifically based methodology for editing and cataloging documents and for generalizing about the characteristics of written language. For a new generation of linguists, ready for another transformation of methodology, such slavish devotion to texts was only part of a narrow, positivistic practice of sacrificing what was alive (the language itself in its oral form) for what was dead and hence more objectifiable (the text itself). This shift in linguistic focus, led by the Swiss linguist Ferdinand de Saussure, both drew upon and in turn abetted the unproblematic model of literacy, for if writing is only a mechanical means of storing living language (rather than an expression of other forms of being), then the study of writing could hardly be of more than passing interest compared with the study of speech. A few decades later, Leonard Bloomfield, one of the central figures in the emergence of modern American linguistics, echoed Saussure's sentiments: "Writing," he wrote, "is not language, but merely a way of recording language by means of visible marks. . . . A language is the same no matter what system of writing may be used to record it, just as a person is the same no matter how you take his picture" (1933:21).

On one hand, Bloomfield's point is incontrovertible. The essential formal properties of language remain unchanged by writing: the utterance, like the subject of a photograph, remains in some key sense unchanged when it is stored, and people interested in studying the essential features of the subject of a photograph would be well served to ignore the representation and to turn directly to the subject itself. On the other hand, however, Bloomfield's distinction lacks interest for anyone concerned with texts not as mechanical reproductions of what already exists (as is the case when speech is transcribed) but as the embodiments of new forms of human expression. In this sense, students of literacy have no more interest in turning from the study of texts to the study of ordinary speech than art historians have in turning their attention from a study of a famous portrait to the study of the model on which it was based.

Of crucial importance for the study of literacy, to use Bloomfield's metaphor, are not the physical differences between a photograph and some original event but the historical differences between what is represented in photographs and what is not—that is, between that which is written down and that which is only spoken. To give just one example, it is important that the intricately plotted detective story did not exist as a genre prior to the ascendency of literacy in the nineteenth century, not that formal distinctions might exist between the written and spoken forms of the same story. Indeed, Bloomfield's metaphor reveals that many texts are not mechanical reproductions at all but works of art, which, like great paintings generally, finally exist on their own terms and must be judged accordingly, even when they are originally drawn from historical models.

As Noam Chomsky makes clear in the following passage, linguistics continues to derive much of its power precisely by eschewing what is most problematic and hence most interesting to the student of literacy:

Wilhelm von Humboldt argued that the resources of a language can be enriched by a great thinker or writer, without any change in the grammar. An individual can expand his facility or the subtlety of his comprehension of the devices of language through his own creative activities or immersion in the cultural wealth of his society. But as in the case of the visual system, it seems quite appropriate to set this matter aside in abstracting the linguistic system as a separate object of study. (1980:234)

Many of the greatest accomplishments of linguistics came about precisely when such concerns with the historical differences between what is different and what is spoken were rejected for the sake of concentration on the formal properties of language. Writing and speaking do share formal linguistic rules in the sense that a Sunday afternoon pickup game and the seventh game of the World Series may share the formal rules of baseball; to the extent that the differences in the rules being used in the two games are in fact minor, or to the extent that each game can lay claim to being a legitimate version of baseball, then someone interested in studying the *rules* of baseball would do just as well observing and analyzing either game. As Martin Schütze noted more than a half century ago, in a critique of the positivistic bent of a language study and the liberal arts generally, "The particulars which are most distinguished and pregnant in their poetic-integral relations are linguistically no more significant than any other verbal data found in contemporaneous speech that has no notable literary relations or value" (1933:282).

Yet what student of baseball would assume that the most compelling difference between the two games would lie at the level of formal rules? On the contrary, the student of baseball, like the student of literacy, would be apt to be interested in precisely the issue that Chomsky dismisses—that is, the "creative activities" that allow individuals operating within the same rules to perform quite differently, indeed in a few cases to perform so splendidly that, rather than restricting their freedom, the rules become a vehicle for self-expression. The lovers of language, like the lovers of baseball, historically turn not to the formal analysis of rules but to literature, to the game itself, to experience the successful playing out of those rules in actual texts that embody the measure of life.

Were it not for the pedagogical and ideological implications of much modern linguistics, matters might be allowed to rest here, with students of language and students of literacy each pursuing what interests them. After all, the informal dinner conversation that lacks certain structural features of an expository essay is still a vital form of communication, just as the sandlot game that lacks all the artistry of the major leagues is still a baseball game—in both cases there is much to recommend the less formal activity either as a pastime or as a subject of serious academic study. All human activities have their own richness and complexity, especially for students interested in diverse forms of expression. The problem arises, however, when the complexity of indigenous practice in general seems so compelling as to make all historical, qualitative comparisons appear either trivial or ethnocentric—as if it were self-evident that, for the astute observer, all language uses have their own richness and complexity. After all, who are we to say that some forms of verbal expression, even the ones that are written down, are any more valuable than others?

Such suspicion of any hierarchical models of communication seems to support the belief that literacy should be defined largely in terms of the actual communicative needs of people within any social setting.

The important thing is not literacy per se but the *functional* needs of real people, whether in the definition of one researcher, as those skills relating to "the ability of each individual in a society to deal with the range of reading and writing tasks that he encounters" (Diehl 1979:9) or, in the words of a Ford Foundation study, more generally as the "possession of skills perceived as necessary by particular persons and groups to fulfill their own self-determined objectives as family and community members, citizens, job holders and members of social, religious or other associations of their own choosing" (Hunter and Harman 1979:9). The notion of "functional literacy" has proven to be so attractive for both language educators and various disciplines concerned with language use in part because it seems to reduce the potentially complex question as to the value of literacy to a level that almost defies analysis: being literate means being able to do in language whatever it is anyone wants or needs to do, or as explained by one radical critic, "Being literate means you can bring your knowledge and your experience to bear on what passes before you" (O'Neil 1970:262).

Literacy consequently has nothing specifically to do with the narrow, perhaps class-based skills necessary to create and comprehend texts; instead it is a generic term that applies to the verbal or even the personal competence of people generally. The ideological appeal of such a notion comes in no small measure from the severing of the traditional ties between literacy and schooling and thus reflects the antieducational, antiinstitutional bias of revisionist thinking. What might otherwise seem to be a series of unusual, almost metaphorical assertions about literacy regularly offered by leading language educators in recent years can be explained in terms of the populist strain in the revisionist critique. For example, Stephen Tchudi advocates a definition of literacy that "acknowledges the literacy of children" (1980:15), and Mina Shaughnessy, in her ground-breaking study of underprepared college freshmen, notes that "the native-born speaker of English has, after all, already absorbed the English language in all its essentials . . . [and] is simply stuck at a number of secondary points where habit does not serve him" (1977:105).

At issue in Tchudi's and Shaughnessy's assertions is the whole question of the actual demands of literacy. On the one hand, it is quite appropriate to downgrade mechanical transcription as a limited accomplishment compared with the unlimited achievement of speech. On the other hand, if one assumes that literacy entails the ability to comprehend and create texts, then it becomes not at all clear what to make of the claim that *all* children, not just the most precocious, are "literate" and that the skills that separate *all* eighteen-year-olds, even the least prepared, from their most accomplished contemporaries, are essentially "secondary points" involving matters of "habits."

While in a sense Shaughnessy is making the important point that most native-born speakers with twelve years of schooling in English should be able to meet the minimal demands of a university writing program, given both a good faith effort on the students' part and suitable academic support services, even this limited assertion must presuppose twelve years of schooling and minimal standards of evaluation. One could just as easily emphasize the immense difficulties entailed in learning how to write well, noting that very few students (and professors as well) can express themselves clearly, forcefully, and fluently in writing despite years of efforts.

Our accomplishments in oral language often have little to do with the difficulties we encounter as writers. Most people clearly do develop the oral language skills necessary to function successfully among family and friends and often at work as well—or they adjust their lives accordingly—yet as Walter Ong notes, "Except for a small corps of highly trained writers, most persons could get into written form few if any of the complicated nuanced meanings they regularly convey orally" (1977:56). While readers attuned to the revisionist critique of literacy may be willing to concede Ong's point, they are, in general, unimpressed by it, for they tend to see the whole exercise of literacy as another form of practice, that is, as a different, and not necessarily better, form of communication. Compared with the relatively minor achievement represented by anyone's learning how to encode and decode language, they reason, a person's mastery of the intricate speech skills that are vital to functioning in the world constitutes literacy in the highest sense. As Tchudi notes, literacy is not an especially complicated concept:

> One doesn't have to set up an elaborate test to determine if another is literate. Rather, answering a very simple question will suffice: Is the person succeeding in doing what he or she wants to do through language? The task may be as simple as ordering a pizza on the telephone or as complex as writing a Ph.D. dissertation. But the test is fundamentally the same: Did it *work?* Did the message arrive? Was the job completed successfully?"(1980:16–17)

But where, we might ask, do these wants really come from? Thomas Sticht, an educator who has written perceptively about literacy in the workplace, takes a different tack in defining functional literacy as the skills needed to perform "some reading task imposed by an external agent" successfully (1975:4). More important in Sticht's view are the literacy tasks imposed by an internal agent, that is, the tasks we impose upon ourselves in the pursuit of a personal goal. Yet even this distinction between internal agent and external agent fails to recognize that we readily call our own many of the things we have to do in life while at the same time denying or being ignorant of what we truly want.

Definitions of "literacy" that avoid any special reference to reading and writing also bypass the critical question of whether or not it is possible for us to formulate wants for things that we cannot imagine. If it is not, what role do reading and writing play in enabling people, first, to imagine other possibilities and other worlds and, second, to make them real? Is it not possible for the ability to comprehend and create texts to be a special ability, unlike any other, precisely in that it enables us to change the world in addition to coping with it? When we define literacy in terms of the skills someone needs to order a pizza over the telephone, we tend to assume that reading and writing are only tools to help people reach goals that are apparent and understandable to them apart from the very acts of reading and writing. While such definitions no doubt have popular appeal, they fail to take into account the special role that literacy can play in helping us, not just to fulfill goals, but to create them as well.

III. Tchudi's definition is very much a part of the revisionist critique of literacy and as such reflects the kinds of concerns underlying many of the central debates about contemporary language education. These issues are perhaps more evident from a comparison of two widely quoted passages:

Passage A

JL ... But, just say there is a God, what color is he? White or black?

LARRY: Well, if it is a God ... I wouldn't know what color, I couldn't say—couldn't nobody say what color he is or really *would* be.

JL But now, jus' suppose there was a God—

LARRY: Unless'n they say ...

JL No, I was jus' sayin' jus' suppose there is a God, would he be white or black?

LARRY: ... He'd be white, man.

JL Why?

LARRY: Why? I'll tell you why. 'Cause the average whitey out there got everything, you dig? And the nigger ain't got shit, y'know? Y'unnerstan'? So—um—for—in order for *that* to happen, you know it ain't no black God that's doin' that bullshit.

Passage B

This is my idea of a Tragic life Dibs a five year old boy, he wasn't understood by his parents the first year of his life was very sad and he din't have any friends, he didn't speak at all, in the classroom he sat by himself.

Passage A is quoted by linguist William Labov to demonstrate the verbal ability of a supposedly poorly educated speaker of a nonstandard dialect. Larry was fifteen years old at the time and a member of a street gang in New York City. Labov's main point is that most people wrongly react to a surface issue, what they feel is Larry's substandard usage, and that they thereby fail to realize that Larry is in fact a "skilled speaker with great 'verbal presence of mind,' who can use the English language expertly for many purposes" (1973:213). Passage B is written by another New York City teenager, here a high school senior, and is quoted by Clifton Fadiman and James Howard as proof "of the trouble we face as a literary society" (1979:9). Somehow, we retain the impression that the respective scholars desire—namely, that Larry is a forceful, even articulate speaker and that the other student is barely educated. In other words, Larry seems to have no language problem, and the other student, no language ability.

It is a mistake, however, to assume that we are dealing with two very different individuals in these two cases. The second student's oral response to the kind of questions asked of Larry might well be as thought provoking as Larry's, and a book report from Larry might well reflect many of the problems apparent in the other student's work. The difference between the two cases likely has to do far less with differences in the young people themselves than with differences in the expectations of Labov and of Fadiman and Howard as to just what kind of verbal skills people should possess. Fadiman and Howard clearly feel that students should be able to write a book report that more closely follows standard usage despite any verbal skills they may possess in other areas: "The liberation of the intelligence (the main goal of literacy) should not be confused with 'socialization,' 'acculturation,' 'self-expression,' or 'the search for identity' " (1979:3).

Labov, on the other hand, just as clearly feels that although the skills Larry possesses may not help him in writing a book report for an English class, they are just as valuable as those being taught in school: "We see," he writes, "no connection between verbal skill at the speech events characteristic of the street culture and success in the classroom; which says something about classrooms rather than a child's language" (1973:207–8). The high school student who can write an "A" book report might not be able to perform the various speech acts that Larry must perform daily in order to survive in the "street culture." For Labov, writing a book report and talking in the street are merely different uses of language; only our cultural bias elevates one at the expense of the other. Behind Labov's support of Larry, therefore, lies a much broader critique of established institutions that see little value in indigenous verbal practice.

But the issues involved here are not as simple as they seem. In citing this passage, Labov demonstrates the fundamental insight upon which the very notion of problematic literacy is based—namely, the recognition that Larry can have symbolic meaning to communicate (here the insight about racial inequity expressed metaphorically by reference to the color of God) without "proper" coding skills, just as other people can have the coding skills and have little to communicate.

The difficulty with Labov's approach, however, is the assumption that Larry's words are equally valuable as speech or writing; to be more exact, the difficulty lies in the confusion he introduces regarding the very status of those words. We have before us, after all, not Larry's speech but only Labov's re-creation of what was said; Larry's speech is here rendered as part of a larger conversation with an interviewer, a dialogue that in turn is embedded in a highly accomplished piece of writing. For Labov to supply the transcription of Larry's words is one thing, but it is a far different matter to supply, first through the active intervention of the interviewer and then through the contextualization of the article itself, the grounds that elevate Larry's words to the level of symbolic text. Here the interviewer plays the role of a trained writing teacher in pushing Larry beyond the limits of synecdochic response— forcing him to verbalize a response about the color of God that objec- tifies rather than merely reflects his position in the world and thus rises above the immediate needs of speech—while the author of the article, here Labov, provides the context that gives this utterance full symbolic status. The anecdote has had an impact on language education in America not because Larry uttered the words but because Labov skillfully used them as a compelling example to support his point in a written text.

Labov's role, in this regard, is not unlike that of a novelist who creates fictional characters whose thoughts and words are eloquent even though they may lack a formal education and may speak a colloquial dialect. While such characters often strike us as realistic, in part because we readily sense that all of us are capable of having such thoughts, their value to us as readers who lack any personal ties to such characters is largely the result of the writer's skills in objectifying the experience of others in texts. We experience, in other words—and that we must value (even unconsciously) in reading any novel—the author's craft as much as the character's speech.

Thus to concede Labov's main point—that Larry does use speech in highly complex ways that both serve his immediate needs and provide a rich study for linguistic and anthropological analysis—is finally to beg a larger question: is there any real difference between complex uses of language in general and complex uses specifically related to literacy? Is there, for example, any fundamental difference between the person who uses language creatively for a personal, immediate purpose and the writer who renders the other person's language for a different, perhaps less immediate, purpose? In answering this question, we must recognize that there will always be speaking situations that even the most skilled writers like Labov would have trouble handling in real life, although they could be re-created in words. To assume that Larry and Labov possess different but fundamentally equal language skills, however, is to overlook the very question that most needs to be addressed—whether or not literacy in fact has any special value.

For every situation in which skilled writers would be at a disadvantage, there are many other situations, including some involving speaking and likely all involving writing, where they have the advantage. A writer as skilled as Labov, for example, can reveal a subject such as Larry to strangers like us in ways that Larry himself probably could not reveal Labov. Indeed, in the absence of literacy, people have little ability to reveal themselves, much less others, to us. Furthermore, it is possible that our understanding of another person as revealed in a text like Labov's, while lacking the richness of that person's self-knowledge gained directly through introspection, may nonetheless have a greater explanatory or predictive power. In the texts of the sociologist, the demographer, the geneticist, and even the poet, there may be important knowledge about Larry to which only skilled readers have access. Furthermore, up to a point these various writers and Larry may have had similar backgrounds, and now these writers may be using their reading and writing skills in an effort to help Larry, perhaps by attempting to organize him and his friends politically as a means of affording them greater opportunities.

One of the main goals in this political activity might be awakening in Larry a more constructive dissatisfaction with the system in which he has learned to function so well. Thus for the political writer, Larry's ability to function well within his world may be as much a part of his problem as the solution. As one commentator of a sociolinguistic study of black adolescent language notes, there is little indication that the young people studied were "in conscious touch with their history," despite the undeniable metaphorical richness of the language: "There is even less to suggest that they have a grasp of the means which ensure their cultural subordination today and which will sentence many of them to a life of economic and political irrelevance" (Mitchell-Kernan 1980:xxii). Literate activity, however, in the form of reading, writing, and general discussion of books and pamphlets, has been one method by which people such as Larry have traditionally been motivated to act in accord with newly discovered interests. Yet precisely at this point Larry's problem merges with that of the student who had such difficulty writing the book report—neither student may be well equipped for the task of comprehending as readers or creating as writers the significant world of experience embodied in literate texts.

For Labov and for many other critics of educational and social policy, literacy, like language, is in general discussed primarily as an instrument for facilitating one's accommodation with society, not for effecting its transformation. The implication underlying such a critique is that the major obstacles to solving one of the most intractable of all humanity's problems, social inequality, have to do largely with the prejudicial attitudes of those in power, particularly as concerns language use. What is needed is less a radical restructuring of either individuals or the historical forces that define their place in the world than some sort of mass enlightenment wherein people would agree to act in accord with moral imperatives about social organization on the basis of certain limited technical claims about language. As a reform-minded educational psychologist notes after reviewing the extensive research refuting the notion of linguistic disadvantage, "If the nonstandard English speaker lived, like some isolated tribe, away from mainstream society, disadvantage would hardly be an issue" (Edwards 1979:97–98). Here the writer might just as well wish that his ahistorical, linguistic account of language activity were an accurate description of the human condition and not the very limit of his methodology. The problem with his findings is seemingly not with the structuralist's model underlying the linguistic research but with a recalcitrant world that refuses to conform to that model.

Or as another reform-minded educator states, "There are culturally different thought processes, as there are culturally different linguistic and paralinguistic processes. All are normal and appropriate in a pluralistic society" (Adler 1979:86). Yet what does it mean to call something "normal and appropriate in a pluralistic society"? To what extent, for example, is the relative inability to read and write texts "normal and appropriate"? Does it mean only that people lacking in such skills should be treated civilly in all personal dealings and that they should be guaranteed equal protection under law? While few would contest either claim, how many would agree that such people are entitled to equal amounts of wealth, status, and power? And even those with the most egalitarian sentiments are still faced with two immensely complex practical problems—how to create and how to maintain this new social order.

At this point the proclamation of linguistic equality rings most hollow, for here the question becomes: what specific language uses are most valuable in promoting such social reform? Educators and other intellectuals adept at literacy can lead the fight, but at what point will those who stand to gain the most from these changes act in their own interest? And when they do attempt to do so, to what extent will their ability to comprehend and create texts aid or hinder their efforts? We need to know, in other words, to what extent literacy is valuable or even necessary in people's efforts to change their historical condition in very immediate, practical ways. If particular uses of language involving the creation and comprehension of texts are necessary or even helpful in this endeavor, then is not this in itself sufficient proof that all language uses are not equal, even in a "pluralistic society," and that all people, even those seemingly content exercising the coping skills they have mastered, should be exposed to the transcending and transforming power of literacy? A skill with the potential of giving one the power to change the world is not just another skill, and we need to learn what all shrewd children know immediately in listening to the classic folktale—that three wishes are unnecessary if we are only wise enough to use one wish to acquire, not finite goods, but the infinite power to remake the world.

On the surface, the critique of literacy has the highest motive of furthering social equality by making us all more conscious of the value of various forms of linguistic practice. Its effect follows this intention by making the literate among us more conscious of the diversity and complexity of verbal communication. The critique of literacy is thus most successful as an act of literacy itself, first as a vehicle for the generation of symbolic understanding by writers and, more important, by making teachers and others more aware of the need for them to change their own perceptions as well as those of their students. We need to be reminded that people like Larry and those who order pizzas by telephone do indeed use language in very complex ways, and as both teachers and researchers, we especially need to be able to recognize the value of all forms of verbal practice.

The conversion of the teaching establishment, however, should not be confused with the direct political action by those whom we teach. The thrust of the revisionist critique is too often directed at the conversion of the teacher, not the transformation of the student, and its moral fervor in no small measure follows the pattern of the pedagogue who is often most passionate within the classroom when most powerless without. We should not allow ourselves, however, to confuse what is more easily controlled (here the attitudes of teachers) with the intransigence of the social world beyond the classroom walls.

IV. Beneath the reformist fervor of the revisionist critics of literacy lies the contradictory impulse of researchers more interested in changing the attitudes of their readers than those of their subjects. Such researchers, represented by social scientists working in the growing field of ethnomethodology, are largely concerned with investigating and revealing the complexity of indigenous practice, as Labov did in his study of the speech of black urban youths. There is, however, an ambivalence at the very center of such research. On the one hand, there is the actual experience of the investigators who devote so much of their professional lives to studying and writing about certain indigenous practices, such as speech patterns of various ethnic groups, in order first to gain a new understanding, in part through the very act of writing, and then to allow others to share that understanding through the act of reading. Here the writers' own commitment both to their research and to changing the perceptions of their readers by disseminating that research in scholarly writing is living proof of literacy's value.

On the other hand, the central theme of such work might well be that the indigenous oral forms of communication under investigation are as demanding and as fulfilling for their practitioners as literate exchange is for readers and writers. In this latter case, the value of a scholar's literate activity lies primarily in its specific content—for example, the insight that certain forms of oral language are highly complex and rewarding—while in the former case the value of the literate activity lies more in the very process of transcending one's own original understanding of the world and then helping others to do the same. Thus what remains unresolved is the ultimate source of such literate activity's value. Is it in what is discovered and hence in the notion that all language use is equal—or is it in the process of discovery and hence in the very practice of literacy? We want to know, by extension, who gains the more, those who actually practice whatever activity is being studied and remain relatively unchanged or those who read and write about it and in the process gain a new understanding of its value.

It is tempting to question the legitimacy of the very distinction being made here between those who engage in indigenous speech practices and those who engage in literacy; each activity is equally valid, if indeed there is any real difference between them. The differences between the two activities, however, do involve important questions of degree. "Participants in social interaction apparently 'understand' many things," writes ethnomethodologist Aaron Cicourel, "even though such matters are not mentioned explicitly" (1973:40). But such matters are mentioned explicitly in the ethnomethodologist's work; indeed, the value of that work lies primarily in making explicit what has heretofore gone unnoticed by everyone and what is likely to remain unnoticed by the participants themselves, despite the new understanding acquired by the person responsible for writing the text and the many people who may read it.

Along with the greater awareness of ethnomethodologists and their readers, moreover, goes a deeper level of symbolic representation. Their understanding of the material embodied in the scholarly text, itself the result of their constructive power of assimilation, represents for them only a *potential* accommodation. Labov's understanding of black urban youths, like Scribner and Cole's understanding of the Vai, is theoretical, revealing to them and in turn to us as readers alternate forms of life that make any subsequent accommodations on our part all the more deliberate. With increased theoretical knowledge on our part (the result of our reading of texts) comes greater freedom for us to choose what we will support and what we will be.

Ethnomethodological critics of literacy resist positing value in this increased freedom of action, since to do otherwise would be to promote a dialectical relationship between the writers of such studies and their subjects—as the writers grew closer to their subjects, the subjects in turn would grow closer to the writers. Such a dialectical relationship would threaten to diminish what had been the original source of value for the writer-researcher in the fundamental difference between themselves and those they study. Quite simply, while ethnomethodological critics welcome the opportunity to remake their readers, they are reluctant to remake the human subjects of their investigation, seeing that the very drive for symbolic understanding that animates their own work is a form of the cultural imperialism that constantly seeks to remake the world in its own image.

One way to curb such imperialistic tendencies is to expand our image of ourselves by learning more about those who are different—and in following such a course, ethnomethodology is unquestionably a literate activity. The difficulty here is that what motivates the writer cannot be allowed to motivate the writer's subjects; people engaged in literacy are expected to change, but the people described in literate texts are not. The dilemma is thus part of what Sartre called "the ambiguity of a discipline in which the questioner, the question, and the questioned are one" (1963:174), even if they are not recognized and treated as such.

Ethnomethodological studies of literacy, for the most part, have not included the "questioned" in the same category as the "questioner." The subjects of such study are often not considered human beings immersed in history and struggling to remake their worlds; instead they are considered natural objects or pieces in a complex structure with which we cannot really communicate. Whereas literate communication can be defined only in terms of a dialectical relationship with a specific synecdochic tradition, one that is constantly changing in response to the persistent symbolic activity of members of the group, the structuralism underlying many studies of indigenous practice considers all communication as part of a complex universal matrix frozen forever in time.

Thus while the student of literacy values any act of communication to the extent that it represents an individual creative effort to transcend the restraints that define our common human condition, the critic of literacy often values any such act mainly for what it reveals about the totality of rules that govern all such communication and that in abstract form compose the total structure of social experience. For such critics inspired by structuralist methodology, there is no dynamic interplay between synecdochic and symbolic communication, between the old and the new; instead all communication, like all social experience, is interpreted as being only a part of an all-encompassing and hence synecdochic whole. The rules themselves, and not any specific, imaginative use of them are the source of human creativity and hence of symbolic experience for the critic of literacy.

"Language," writes the anthropologist Claude Lévi-Strauss, "presents us with a dialectical and totalizing entity but one outside (or beneath) consciousness and will. Language, an unreflecting totalization, is human reason which has its resources and of which man knows nothing" (1966:252). The critics' concerns, then, are neither with their own experiences nor with those of the subjects they study but only with the larger pattern of behavior that all humans exhibit, which in turn is interpreted by the structuralist methodology as expressing a meaning about humanity freed from the contingencies of history. The truth at the very heart of literacy that we seek to learn from studying the responses of others to the experience of the contingency that is our common historical condition Lévi-Strauss, for one, dismisses from the structuralist's perspective as an "overindulgent attitude towards the illusions of subjectivity":

The raising of personal preoccupations to the dignity of philosophical problems is far too likely to lead to a sort of shop-girl metaphysics, which may be pardonable as a didactic method but is extremely dangerous if it allows people to play fast-and-loose with the mission incumbent on philosophy until science becomes strong enough to replace it: that is, to understand being in relationship to itself and not in relation to myself. (1974:58)

In discounting the dialectical relationship between writers and subjects of study, even when both are human, Lévi-Strauss creates the illusion of studying, not purposeful human activity, but an agentless natural world.

In the revisionist critique, reading and writing are no longer acts of praxis, creative attempts to reshape the world as a means of reshaping the self. Rather, they begin to assume some of the same ritualistic elements that they have among people who possess little constructive assimilative power. Without a realm of social interaction and human development, there is little place for the immediate practical connection between thought and action that lies at the very heart of literacy. To this extent, it is possible to see the promulgation of structuralist-based critiques of literacy as an analysis and also a reflection of the perilous condition of literacy in our age. As social theorist Mark Poster writes, "With the unchecked growth of bureaucratic structures in advanced industrial societies (structures that were ruled by no one), with the dissolution of historical consciousness (the sense that the future was not in the hands of the people), it should not be too surprising that a theory developed in which society was composed of agentless structures" (1975:317–18). Or as Fredric Jameson notes, "There is therefore a profound consonance between linguistics as a method and that systematized and disembodied nightmare which is our culture today" (1972:ix).

At first there may seem to be little connection between the cultural despondency and the reformist fervor, seemingly so opposite in nature, that lie at the heart of the critique of literacy. These two conditions are nevertheless different faces of the same cultural condition, one characterized by the collapse in the belief in developmental patterns of symbolic representation like the one outlined by Piaget and, in turn, by the collapse in the belief that there is some vital connection between our control of the world and our understanding of it. Instead, we feel ourselves moving in either of two seemingly antithetical directions— either toward a greater scientific knowledge of a world we can study but never control (and hence toward a sense of *objectivity* that is increasingly remote) or toward a greater personal involvement with our own feelings (and hence toward a sense of *subjectivity* that is increasingly personal).

The totality of social experience is part of a distant objective world, while the totality of our own feelings is part of an intimate subjective world. The prevalence of such a condition leads to the widespread contemporary belief that real knowledge of the world must be "objective" just as real knowledge of ourselves must be "subjective." As a consequence, we find ourselves alternately studying the world or changing ourselves—the one thing we seem unable to do, or often even to consider, is changing ourselves as a means of changing the world.

The confusion about the proper role of theory and in turn the confusion about the nature of objectivity and subjectivity can also be seen in certain aspects of reader-response criticism, especially in its concern with unmasking what it sees as the ideological basis of reading. For David Bleich, for example, the fundamental task of criticism is to recognize the subjectivity that underlies all reading, even the seemingly objective readings of authorities. In making his case, Bleich first establishes a clear distinction between "objectivity" and "subjectivity" based strictly upon whether or not an observer is involved: "When the observer is part of the observed truth is no longer objective" (1975:752). Then he charges that the promotion of any such observed truth as *objective* is really a disguised form of cultural domination: "The notion of objective truth has the same epistemological status as God: it is an invented frame of reference aimed at maintaining prevailing social practices" (1976:317).

What strikes us as having the authority of "objectivity," in other words, is likely to be only the ideas, opinions, and beliefs of the prevailing ideology. "Objective" knowledge, therefore, is really a means of legitimating existing power relations. The difference between objective and subjective is grounded in existing social relations, not in categories of truth. Bleich here follows the structuralist model in associating objectivity with the realm of knowledge that is unaffected by the activity of the viewer. All other forms of knowledge, including all those that involve a dialectical relationship between human beings, are thus labeled "subjective." All knowledge derived from reading, as well as all historical knowledge, is thus inherently subjective.

Yet to acknowledge the historical basis of all understanding is to condemn us all to subjectivity only if we assume that our historical conditions are a fixed entity that remains unaffected and hence equally corruptive of objectivity, regardless of our most strenuous efforts at self-understanding. It is self-evident that we can only know the world from our own point of view, but it is not self-evident why we must assume that this point of view is somehow permanently fixed, forever limited, despite all our efforts, most notably including our efforts to comprehend and create symbolic meanings in texts.

The concerns of reader-response criticism are misdirected in locating the objectivity of the text, like the objectivity of history, in some straw figure drawn from natural science, a figure that forever divorces subject and object even when both are human. It is mistaken to assume that objectivity must somehow exist in the world of objects, free of all human intervention, with the implication that all other experiences of the world are therefore subjective. On the contrary, objectivity, to the extent that it describes a condition of human knowledge, exists only within our ever-deepening acts of symbolic representation; that is, objectivity exists only in our efforts to free ourselves from the moment by forming more deliberate accommodations and more constructive assimilations.

All understanding is grounded in the experience of an individual, in what Bleich calls subjectivity, just as all representational activity is itself grounded in imitation and play, but in both cases what matters most is the deepening of the interactive process itself, which transforms the world as it transforms ourselves. Our goal, therefore, is less to escape subjectivity than to transform it and, in so acting, to follow the dictate of Heidegger's advice about dealing with the circularity of human understanding: "What is decisive is not to get out of the circle but to come into it the right way. The circle of understanding . . . is not to be reduced to the level of vicious circle, or even a circle which is merely tolerated. In the circle is hidden a positive possibility of the most primordial kind of knowing" (1978:195).

Like Heidegger, Bleich is interested in revealing the individual nature of all understanding. "An observer is a subject," he writes, "and his means of perception define the essence of the object and even its existence to begin with" (1976:318–19). His interest, however, unlike Heidegger's, is not in seeing individual understanding as the basis of our historical condition and hence the source of all knowledge. For Bleich, objective knowledge still exists; what is important for individuals is that they not confuse objective knowledge with the inherently subjective readings of authorities. Bleich's goal, in other words, is not to ground all knowledge in historical experience but to demolish the myth that the readings of authorities constitute *objective* meaning and in its place to erect a model that democratizes the reading process by recognizing its fundamental *subjectivity*. All readers have their own subjective response, which, given the subjectivity of all responses, including those of so-called authorities, has its own value.

Instead of aiming for a convergence with the objectivity of the text, along the lines of Gadamer's metaphor of a "fusion of horizons," Bleich stresses the need for an awareness of one's subjectivity: "The classroom aim is to teach students to develop their own subjective judgment and to be responsible for it" (1976:333). Here the difference between the symbolic meaning of the text and the synecdochic status of opinion is replaced by a new moral imperative. In such a world, the truth or falsity of a belief is apt to become less important than the willingness of individuals to accept personal responsibility for their beliefs.

As Lévi-Strauss so poetically observes in *Tristes Tropiques*, the source of literacy and, in turn, of human knowledge itself, is our dissatisfaction with the man-made world we inherit, a world that exists for us, not as our creation (although it was created by other humans), but only as our condition. We are all capable of setting out like the young Lévi-Strauss to change our understanding of the world and thus, in a limited sense, the world itself. Critics of literacy err fundamentally in confusing *culture* (and hence significant human achievement) with that which we inherit rather than with that which we create. Culture should not be equated with our condition—the artifacts and ideas that surround us (and indeed often imprison us). Rather, culture needs to be identified with the active spirit within us that first created this world and is continually seeking to remake it, in no small measure through the written records of our acts of interpretation.

The spirit that leads us to remake ourselves and our readers in our own literate activity is the spirit of culture; while we may dread contaminating others with our culture, we should never act directly for others, shielding them as it were from our world. We can study others as if they were natural objects, but ultimately our common humanity insists that their value to us as objects must never surpass their value to us as subjects capable of acting for themselves and of studying us just as we study them. If people must change to meet this condition, if they must lose something of what we had valued so highly in them originally—quite simply, if they must become more like us—then so be it, for at the same time, in our very act of understanding, we are becoming more like them.

There are many different life-styles and forms of social organization, and in studying them we necessarily grow by incorporating the other into ourselves. The forms of social organization can be protected and preserved for future generations, but the spirit that gives true value to all cultural experience is much more fragile. We can preserve the spirit of culture only when we encourage people everywhere to form more deliberate accommodations based upon more constructive assimilations, that is, when we encourage human development. Culture, like literacy, exists, not in what we inherit, but only in what we create, not in the rules that govern our activities, but only in our imaginative playing out of those rules. Culture, writes social theorist Zygmunt Bauman, "is the most audacious of all attempts to scrap the fetters of adaptation. . . . [It] is a daring dash for freedom *from* necessity and freedom *to* create. It is a blunt refusal to the offer of a secure animal life. It is—to paraphrase Santayana—a knife with its sharp edge pressed continuously against the future" (1973:172). The critics of literacy can wield this knife like a surgeon's scalpel; they are reluctant, however, to trust others with such a weapon, preferring instead to convince us all that we may soon live in a peaceable kingdom.

6

Literacy and Social Reproduction

I.   Our sense of the value of literacy derives in part from our belief in social progress; learning to read and write, accordingly, encourages us to break free from traditional patterns of social reproduction. Illiteracy and ignorance are the chief obstacles to progress, just as literacy and education are its chief vehicles. With the Enlightenment came this notion of the liberating, progressive nature of education, arousing the hopes of thinkers like James Mill that somehow universal literacy would prove to be the salvation of humanity. As John Stuart Mill wrote, his father "felt as if all would be gained if the whole population were taught to read, if all sorts of opinions were allowed to be addressed to them by word and in writing, and if, by means of suffrage they could nominate a legislature to give effect to the opinions they adopted" (1924:74). The free exchange of information promoted by universal literacy will thus result in the uniform progress both of individuals and of society generally. Today, this ideal enlivens the reforms of the Brazilian educator Paulo Freire and, in this country, of Jonathan Kozol, two writers who see the extension of literacy as a crucial step in the political and psychological liberation of those who have hitherto been denied power.

Similarly, conservative forces initially opposed many of the efforts to expand literacy and schooling, seeing such actions as a threat to the existing social order. Throughout Europe in the eighteenth and nineteenth centuries, there were frequent discussions of the dangers posed by teaching the lower classes to read, since access to books raised the possibility that people might seriously consider other ways of earning their living or of organizing society generally. In England, for example, the president of the Royal Society led a successful fight in the House of Lords against a bill to establish a national system of elementary schools precisely because he sensed that the symbolic dimension of literacy was an inherent threat to indigenous practice:

> However specious in theory the project might be of giving education
> to the labouring classes of the poor, it would in effect be found to

be prejudicial to their morals and happiness; it would teach them to despise their lot in life, instead of making them good servants in agriculture, and other laborious employment to which their rank in society had destined them; instead of teaching them subordination, it would render them factious and refractory, as was evident in the manufacturing counties; it would enable them to read seditious pamphlets, vicious books, and publications against Christianity; it would render them insolent to their superiors; and in a few years the result would be that the legislature would find it necessary to direct the strong arm of power toward them. (Quoted in Cipolla 1980:65–66)

American slaveholders certainly understood what was at stake, passing antiliteracy laws in South Carolina and Georgia as early as the middle of the eighteenth century. "Everything must be interdicted," wrote a justice on the Georgia Supreme Court, "which is calculated to render the slave discontented" (Genovese 1974:562). As Genovese notes, despite the specific reasons slave states may have had for outlawing literacy, their general concern was evident: the ability to read was accompanied by greater insight into one's condition in the world and hence greater potential for purposeful action, even for those without political power, such as former slave Josiah Henson:

It was, and has been ever since, a great comfort to me to have made this acquisition; though it has made me comprehend better the terrible abyss of ignorance in which I had been plunged all my previous life. It made me also feel more deeply and bitterly the oppression under which I had toiled and groaned; but the crushing and cruel nature of which I had not appreciated, till I found out, in some slight degree, from what I had been debarred. At the same time it made me more anxious than before to do something for the rescue and the elevation of those who were suffering the same evils I had endured, and who did not know how degraded and ignorant they really were. [(1849) 1970:137]

Quite simply, people who could read would no longer want to do and believe what they were told.

Such a belief in the direct connection between literacy and the reform of the social order is supported not only by the problematic nature of literacy (with its notion that the content of literate communication always entails going beyond the limits of ordinary speech) but by the symbolic nature of schooling as well. Schools tend to separate learning from practice, and their mere existence, as Bruner observes, is an indication that there is "knowledge and skill in the culture far in excess of what any one individual knows" (1973:12). Even Sylvia Scribner and Michael Cole, who are suspicious of the exaggerated claims about the cognitive power of literacy, readily admit that schools do work to loosen the grip of tradition by heightening people's awareness both of alternative forms of practice and of their own mental operations. Learning that takes place within the structure of everyday life does tend to be more traditional in stressing the continuity between generations; learning that takes place within schools, however, follows the pattern identified by Margaret Mead as placing "a heavy emphasis upon the function of education to create discontinuities" (1943:637).

Even though both schooling and literacy seem to be inherent forms of praxis, it is perhaps far more accurate to see them only as *potential* forms, that is, only as acts of praxis in theory. It is mistaken to see either children's attendance in schools or their reading of books as somehow automatically involving a transcendence of synecdochic practice. While the process of schooling, like the process of reading, does allow for symbolic response, the very acts themselves can just as readily take the form of indoctrination in the service of prevailing attitudes. Many of the traditional functions of socialization, in other words, can easily be transferred to the school and can with little difficulty be incorporated into the reading and writing curriculum.

As demonstrated in chapter 2, students learned reading skills and, to a lesser extent, writing skills in schools for hundreds of years before much interest was shown in the symbolic nature of that which was being communicated. Even today, many educational practices and "reforms" reflect the efforts of various groups to produce students concerned less with pursuing symbolic meaning in acts of praxis than with thinking and acting in ways that reflect some preconceived notions of correctness. Thus despite whatever liberating potential either schooling or literacy may have, there are still powerful social forces at work concerned largely either with teaching students the right responses to questions or with giving them practical training in the performance of some particular set of tasks.

Within any society, schools have always fulfilled diverse functions, some of which have little to do with promoting literacy, while others may have been openly opposed to it. Drawing upon Max Weber's tripartite division of social groups into those concerned primarily with class, status, or party, Randall Collins (1977) locates three encompassing and often competing demands placed on education. Accordingly, schools serve to train people in specific, practical skills, to promote the social standing of various groups, and to ensure political stability. For most of history, Collins notes, schools have played only a minimal role either in instructing people in practical skills, since such skills were usually taught at home or through apprenticeships, or in guaranteeing political stability, since this role was more effectively played by the more pervasive institutions of the church and the family.

The essential function of schools historically was therefore to promote the status of elites. Even now, when the mastery of minimal skills in coding and computation seems so necessary, the evidence gathered by Thomas Sticht suggests that much of the actual training for work still takes place on the job through direct, mimetic instruction, with little dependence upon the comprehension of texts.

The state-sponsored, compulsory education that sprang up in practically all industrial countries in the second half of the nineteenth century, Collins argues, responded less to the need of industries for workers with higher skills than to the political needs of the state itself to exercise more effective control over newly mobile populations. People in the nineteenth century were highly employable when they entered the work force with only minimal coding skills; if this is no longer the case today, we should hardly be surprised that the actual level of student attainment remains so low or that the curriculum and pedagogy now being used still tend largely to reflect the reproductive role of schooling in enforcing existing beliefs and practices.

Between the ideal nature of schooling as training in symbolic thinking and the actual pedagogic practice that takes place within the typical classroom, there is indeed a wide gap. One indication of the dimension of this problem is a recent report on the amount of writing that students really do in American high schools. In the study *Writing in the Secondary School,* Arthur Applebee reports that, while an average of 44 percent of the class time observed in the research involved students in some form of writing, only very little of this writing was concerned with creating texts. Over half of the class time spent on writing was devoted to what Applebee calls mechanical uses of writing (what might be called "writing without composing"), including such tasks as completing multiple-choice and fill-in-the-blank exercises, and one- or two-sentence responses to questions. Of the remaining time spent on writing, nearly all was occupied with note-taking. Overall, therefore, only 3 percent of the total time devoted to a class, including homework, required students to produce at least a paragraph of coherent text. And even these paragraphs tended by and large to deal with reporting and summarizing accepted opinion. Less than one-half of 1 percent of the work done in contemporary American high schools studied in this comprehensive research project involved students' doing writing that required them to express their personal response.

Even in high school English classes, in Applebee's terms the "traditional center for writing instruction," only about 10 percent of class time required students to write coherent paragraphs. Given the history of writing education as we know it, Applebee might perhaps more accurately have called the English classroom the "center for traditional writing instruction," that is, the center for instruction in correctness in the mechanics of written language, including spelling, punctuation, usage, and grammar. The English classroom has been the center of such language instruction in high schools, at times dubbing it "grammar," now more often "linguistics," just as it has been the center for the study of literature; however, we may well question the extent to which the study of language translates into the ability to create texts or the study of literary history into the ability to comprehend them. We have no assurances, in other words, that even the actual practice of English in contemporary American high schools has today or has ever had as its primary goal making students literate readers and writers of texts.

Thus while schools generally and literacy education specifically seem ideally to support the systematic reform of society, in actual practice schooling, including much of the current reading and writing curriculum, seems to support systematic social reproduction. As Max Weber (1958a) noted in his discussion of the Chinese literati, reading and writing can be made particularly difficult and far removed from synecdochic exchange as a means of enhancing the status of the ruling class and the bureaucracy that supports it. Rather than following the lead of Freire and Kozol, most contemporary radical critics of education have tended to emphasize the fundamentally conservative nation of schooling and, as part of their critique, have tended to see language instruction less as a vehicle for social progress and more as an obstacle to it. Certain aspects of writing instruction—for example, the emphasis on adherence to standard forms of usage—are thus often pictured as a means of promoting the style and the values of the ruling class under the guise of educational achievement.

What remains uncertain in this critique, however, is not the generally conservative nature of schooling as it is actually practiced (about which there can be general agreement) but the political dimension of literacy itself. In other words, the extent to which literacy education plays a *necessarily* conservative role in the process of social reproduction remains uncertain. Learning how to read and write may invariably be a form of socialization into the "dialect" and overall value system of the dominant class, or perhaps only the current reading and writing instruction, based as it largely is upon a limited model of literacy, is so conservative. In other words, in learning how to code language properly, we may possibly be engaged only in social reproduction, while in learning how to create and comprehend texts, we may be engaged in social reconstruction.

To the extent that literacy education is itself necessarily conservative, then reform involves either devaluating or democratizing it so that a diversity of models may be constituted as equally legitimate; however, to the extent that only the current pedagogic practice is so conservative and not literacy itself, then the efforts at reform should change the focus of reading and writing instruction in order to ensure that everyone does eventually have an opportunity to learn how to comprehend and create texts. Without resolving this basic issue it is impossible to understand, much less to change, the social basis of literacy education.

II. Critiques of literacy often reflect more comprehensive critiques of the cultural domination of the ruling class. According to such a critique, in promoting the value of literacy, the dominant class is really promoting its own attitudes and accomplishments as absolute standards, in the process using institutions like the school to measure the relative inabilities of other groups to meet these standards. "In every epoch," Marx writes, "the ideas of the ruling class are ruling ideas, that is, the class that is the ruling material power of society is at the same time its ruling intellectual power" (1967b;438). Societies generally and schools specifically demand from all students only what select students, based upon their class background, can readily produce. Thus schools are an especially insidious means of social reproduction, since, on the one hand, they disrupt indigenous practice and in so doing seem to bring the power of symbolic representation to all equally and, on the other hand, they are so constituted that those whose socialization at home makes them on the whole more at ease with symbolic exchange will likely succeed in school at the expense of those who are ill prepared to deal either with verbal meanings or with experiences generally out of context.

Schools do take students from the synecdochic social world and place them in the symbolic arena of the classroom, and the content of instruction is not exclusively the actual practice of any one group (and thus something that would clearly give that group an unfair advantage) but either knowledge that exists independent of any one group (for example, physics) or the abstract skills necessary to deal with such knowledge (for example, mathematics). Accordingly, the most important element in symbolic education is not what one has already learned but one's capacity to learn something new. It is no accident, therefore, that the emergence of mass schooling and problematic literacy in the late nineteenth and early twentieth centuries (both concerned at least nominally with symbolic negotiation) was accompanied by the rise in intelligence testing designed to measure the human cognitive capacity to negotiate symbolic meaning—a capacity that some at the time felt existed independent of the synecdochic constraints of socialization.

This testing movement undoubtedly had a conservative component, one concerned with establishing "scientifically" the natural superiority of the dominant classes and the dominant race. The movement also had a reform component, however, since, like schooling and literacy, testing seemed to offer another means of exchanging the traditional mode of reproduction, wherein people's positions in the world were largely determined at birth by the social standing of their parents, for one that would reward people and thus reshape society on the basis of innate skills. The scientific measurement of people's capacity to think, as compared with the measurement of what they had actually learned, thus did seem to offer the hope of an entirely new mode of social reproduction—one leading to a meritocracy in which those more adept at symbolic negotiation would rise to the top, regardless of their social and ethnic origin.

In the last twenty years, revisionist historical critiques such as those of Harvey Graff have exploded the idea that schooling and literacy, taken either separately or in combination, are as efficient in promoting social mobility as reformers had anticipated. Such revisionist studies demonstrate that allocating rewards on the basis of people's ability to negotiate symbolically in "intelligence" tests or in tests of reading and writing in no way results in a meritocracy—unless, that is, one adopts a theory of racial or genetic inferiority to explain the fact that for the most part the children of the poor are not as adept at symbolic negotiation as the children of the rich. If the model of the meritocracy operates at all within our society, it does so only in very limited areas, for example, in certain sports like basketball, where the rewards for excellence are great and where the attainment of the necessary skills does not ordinarily require costly coaching or equipment at an early age. In most other areas of society, however, even in other sports such as tennis (where the prerequisite skills usually do require prolonged instruction, in part because they are less imitative and more highly constructive in a Piagetian sense), social opportunity seems to play a far greater role in determining who will be successful.

Thus the first key question to consider with regard to literacy and social reproduction is whether the negotiation of symbolic verbal meaning more greatly resembles the jumping skills required in basketball or the racket skills required in tennis. The answer here is hardly in dispute, since by definition symbolic negotiation depends upon the socially nurtured inclination to break with established practice. The crucial role of socialization can be seen, for example, in one of the common tests of classification used to measure "intelligence." Children are asked what two objects such as an apple and an orange have in common, and the scoring of the response is based upon the notion that some answers ("both are eaten") are more imitative and thus less constructive than other answers ("both are fruits"). Quite likely, children from all social backgrounds, regardless of their actual responses, know that an apple and an orange are both fruits, just as they know that both can be eaten. What does require explanation, however, is the social basis underlying the pattern of response that generally elicits synecdochic or symbolic responses from children that are based largely on economic standing and thus in turn on a larger pattern of social reproduction. We should be able to explain, in other words, why class affects such acts of classification.

Some critics would intervene here to dismiss the whole distinction between synecdochic and symbolic response, arguing that this very distinction is a form of ethnographic domination, playing up the indigenous practice of the more powerful group by attributing to it some special, ill-defined *symbolic* quality. A "symbolic" response, in other words, is merely the "synecdochic" response of the ruling class. Classifying an apple and an orange as two fruits rather than as something we eat is only another, subtler form of imitation. Yet as Basil Bernstein's research (1981) into the social basis of classification indicates, such a response ignores a fundamental point: children who are apt to give initially the superficially less imitative response (for example, that cheese and milk have a similar origin) are equally adept at forming the apparently more imitative response (for example, that both are eaten at lunch).

For some reason, however, they interpret the request from an examiner for *any* classification as a disguised request that they demonstrate their ability at symbolic negotiation by responding with the least imitative grouping. Thus on such simple classification tests children and people generally seem to be differentiated not by some reified entity called "intelligence" or even by some more locally defined skill but by the presence or absence of an implicit, psychologically internalized sense of the higher value of less imitative, more constructive representation. Groups of people, defined largely by their actual social experiences, seem to be more or less motivated to produce such symbolic responses.

In one of his early essays, Bernstein sums up the fundamental connection between social life and what he calls a "communication code" (perhaps better described as the verbal manifestation of this psychological predisposition toward symbolic discourse):

The genes of social class may well be carried less through a genetic code, and far more through a communication code that class itself promotes. This communication code emphasizes, verbally, the communal rather than the individual; the concrete rather than the abstract; substance rather than process; the here and now rather than regulated future; consequence of action rather than intent; and positional rather than personal forms of control. To say this about the communication system is not to disvalue it; such a communication system has a vast potential, a vast metaphoric range, and an aesthetic capability. A whole range of meanings can be generated by such a communication system. It just happens not to be very helpful for success at school or for social mobility. (1967:18)

The ability to generate and comprehend symbolic verbal meanings, in other words, is as deeply embedded in a person's social condition as the ability to play professional tennis. Furthermore, although the natural skills required to perform the activities in both cases are likely universally present in all people and are thus evenly distributed across class lines, the actual development of these skills depends largely upon historical circumstances that shape the future for all but the rarest individuals. Public tennis courts are available to all, just like public libraries, but it seems pointless to equate this highly limited equality of access with full equality of opportunity. Bernstein's point is that, when families generate wealth more by means of powerful, constructive assimilations and less by means of strict accommodation, the children are more likely to be encouraged to generate and comprehend, orally as well as visually, meanings that are far removed from the immediate, practical concerns of social life, and success at negotiating such meanings best explains the relative overall success of such children in school as well as in "intelligence" tests and in standardized tests of reading and writing.

Critics of Bernstein have been quick to note that the preference for one kind of communication rather than another resembles the preference for one game or one style rather than another. It is as if the dominant class had chosen to reward tennis players rather than basketball players, knowing all along that in doing so it would gain the advantage of seeming to offer equal opportunity to all ("The best players *regardless of background* will win the prizes!") while in fact it guaranteed that most of the rewards would accrue to players whose families were able to give them the economic and psychological support necessary to develop championship skills (particularly the desire to succeed at an individual game rather than a team game). Similarly, the school can be viewed as an arena to which all come to compete, seemingly on equal grounds, but in which the selection of what is and is not to be judged and hence rewarded largely predetermines that certain groups will receive a disproportionate share of the rewards.

French sociologist Pierre Bourdieu, working with Jean-Claude Passeron, sees schools as the perfect instrument for legitimatizing authority in a democratic state precisely because they appear to give everyone an equal chance to succeed while systematically and efficiently reproducing the social order:

In ever more completely delegating the power of selection to the academic institution, the privileged classes are able to appear to be surrendering to a perfectly neutral authority the power of transmitting power from one generation to another, and thus to be renouncing the arbitrary privilege of the hereditary transmission of privileges. . . . The mobility of individuals, far from being incompatible with reproduction of the structure of class relations, can help to conserve that structure, by guaranteeing social stability through the controlled selection of a limited number of individuals—modified in and for individual upgrading—and so giving credibility to the ideology of social mobility whose most accomplished expression is the school ideology of *"l'Ecole libératrice,"* the school as a liberating force. (1977:167)

Schools are particularly effective, Bourdieu believes, in convincing students that they in fact had an equal chance and thus that their failure is their own fault. In this sense, schools legitimate the inequality of the social world in the minds of those who are most discriminated against, creating what Bourdieu calls "dependence through independence."

Where schools are concerned Bourdieu is mainly interested in French university life, and he often has in mind the characteristic aloofness of French professors and a curriculum that values style—which can be learned only with the greatest difficulty apart from the whole process of socialization—to content. Thus in Bourdieu's European view of education, the Jay Gatsbys of the world are condemned to be outsiders and failures in a system geared to reward not the outward signs of success that can be readily purchased but the inner grace, the manner that often proves all the more elusive to those most desirous of it. Culture itself is valued less than, in Bourdieu's words, "the relation to culture"—all things that set "against the vulgarity of what can be acquired or achieved a manner of possessing an acquirement whose value derives from the fact that there is but one way of acquiring it" (1977:130), that is, indirectly, through the process of socialization itself.

Like Bernstein, Bourdieu sees this fundamental relationship to life as playing itself out in people's actual use of language; just as one has a "relationship to culture," one also has a "relationship to language." The aloofness and paraphrasis of professorial language, in other words, is merely an extension of the class standing of the professors themselves, and once again, the crucial element is one's ability to transcend immediate social context.

The disposition to express feelings and judgments in words, which is greater the higher the level in the social hierarchy, is only one dimension of the disposition demanded more and more the higher the level in the educational hierarchy and the hierarchy of occupations, to manifest, even in one's practice, the capacity to stand aloof from one's practice and from the rule governing that practice. Contrary to appearances, nothing is more opposed to literary ellipsis or metaphor, which almost

always presupposes the context of a literate tradition [that is, a familiarity with the symbolic world of texts], than the practical metaphors
and "ellipsis by *deixis*" . . . which enable working-class speech to supply
all or part of the verbal information by implicit (or gestural) reference
to the situation and "circumstances." (1977:117)

All aspects of language use, Bourdieu concludes, from accent to phrase-
ology to rhetorical devices, "express" the choices of the speaker less
than they "betray" those choices, revealing in the process what Bourdieu
calls "a relation to language which is common to a whole category of
speakers because it is the product of the social conditions of the
acquisition and use of language" (1977:117).

For Bourdieu, the school is not really an active participant in social
reproduction, not an instrument of the dominant class, as more radical
critics contend. Rather, it is endowed with legitimacy by that class
when it fulfills its own liberal, scholarly pursuit of knowledge. Schools,
in other words, are very effective agents of reproduction because their
principal concern is not reproduction but "truth" and academic excellence. The ideal of disinterested scholarship represents the ideological
triumph of the professional, upper middle class. What remains unclear
in Bourdieu, however, is the extent to which matters could be altered
in an effort to promote greater social mobility between generations.

In part this matter is obscure because Bourdieu's interest seems to
be largely, if not entirely, in unmasking the ideological, legitimatizing
function of schools. And on this general point, there is little room for
disagreement—schools do play a central role in social reproduction;
indeed, only in the rarest circumstances, and then perhaps only for
brief periods of time, are they solely or even principally agents for a
radical restructuring of society. Yet part of the reason, as I have already
noted, is that, in practice, schools tend to be far more concerned with
promoting the status of various groups and with ensuring political
stability than with teaching symbolic negotiation. We therefore cannot
determine from Bourdieu's critique whether literacy education has the
potential for promoting greater social transformation or whether it is
necessarily only another means of enhancing the status of the ruling
class and thus necessarily only another form of social reproduction.

Is indirection an important element in academic discourse, we want
to know, because precisely this element is lacking in working-class
speech and thus ensures the failure of working-class students in comparison to other students? In other words, is indirection merely an
affectation of the dominant class and hence chosen almost gratuitously,
or is indirection important for other, perhaps more valid reasons (for
example, because as a limited symbolic response it indicates a person's
more general ability at the kind of symbolic representation that, at the

very least, is valuable economically in a given historical situation)? In highlighting the artificiality of academic discourse, Bourdieu suggests that the only thing at stake is a series of verbal conventions that are valuable only to the extent that those with power possess them and those without power do not—and could not readily acquire them, even if they tried.

In noting that one's "relation to language" is a social product, not a linguistic or genetic one, however, Bourdieu raises the possibility that differing relations to language are merely part of a larger, more encompassing pattern that is rooted in a concrete historical situation. In this regard, Bourdieu notes that people who lack the ability to use language indirectly, in part because their whole relation to culture is more direct, are also likely to lack an adequate understanding of their own plight— are also, in other words, likely to believe that the system is fair and to blame themselves for their failure. Schools, writes Bourdieu, manage to "convince the disinherited that they owe their scholastic and social destiny to their lack of gifts or merits, because in matters of culture absolute dispossession excludes awareness of being dispossessed." In taking such a tack, Bourdieu seems to imply both positions—namely, that certain patterns of language use related to literacy (although perhaps not literacy itself) are gratuitously chosen to ensure reproduction and that individuals who lack these skills are in fact disadvantaged in not being able to understand fully what is happening to them.

For the most part critics of literacy prefer the notion that what is at stake here is essentially the arbitrary preference for one style or mode of verbal expression rather than some other different but equal form of expression that happens to be indigenous to a group with less power or prestige. To the extent that such a critique is directed toward schooling generally (principally concerned, as it is, with status and party) and not specifically to literacy, there seems to be little to disagree with. A great deal of schooling, including many aspects of current language education, is concerned with social reproduction and clearly does overvalue gratuitously chosen characteristics of the life-styles and beliefs of those in positions of power. Patterns of pronunciation certainly fall into this category, as do distinctive verb forms and idioms generally.

It is less certain, however, whether literacy itself, as the ability to comprehend and create symbolic texts, is merely another means of cultural domination; part of the difficulty in resolving this issue is the fact that so much of the critique of literacy already considered is really directed against the unproblematic model of literacy and hence against what seems to be a patently ethnocentric overvaluation of the ability to encode and decode oral language. When such criticism is directed against education generally, the problem is only compounded as the issue of students' ability to comprehend and create texts almost falls out of the picture entirely.

The academic content of the curriculum generally, not just that in reading and writing, does seem to be far less important for analyzing the social function of schooling than is the teaching of the hidden curriculum in which students learn to comply with a vast array of written and unwritten rules. And when direct instruction in reading or writing is considered, attention generally focuses on the various ways in which the curriculum concerns itself with instruction in the most superficial aspects of unproblematic literacy.

In *Schooling in Capitalist America,* for example, Samuel Bowles and Herbert Gintis basically ignore the question of whether teaching students to read and write books, even ones like their own, promotes or hinders the radical transformation of society they so ardently seek. Their concern, like Bourdieu's, is to expose the vast disparity between the potential of schooling to transform individuals and society and the actual practice of American schools during the nineteenth and twentieth centuries, a practice that they claim has led to the reproduction of a social order responsive to the needs of a capitalist economy. Compared with the school's theoretical potential for liberation, so forcefully articulated by educational reformers, their actual achievements in restructuring society does appear disappointing. The fundamental question that remains unanswered, however, is what to do about this situation. How we are to go about changing society, and what role should the teaching of literacy in schools play in this effort? These questions remain unaddressed.

Believing that the content of all schooling is essentially arbitrary, Bowles and Gintis cannot view the commitment to mastering knowledge divorced from context as anything but a subtle scheme to achieve cultural domination. They advocate instead that schools should become arenas, indeed training grounds, for the very struggles that dominate social life. The people who are oppressed in society should fight to gain control of the schools, both as a direct means of gaining power and as a means of training for the larger social struggle that must take place outside the schools. The schools will therefore become models for the sharing of power in a truly democratic society.

Bowles and Gintis, however, can give only the most general sense of how these changes are to be brought about. After a couple hundred pages of very specific criticism of the reproductive nature of schooling, they are able to offer only three rather modest suggestions for reform: first, that we recognize the struggle for social equality as a political struggle, not an educational one; second, that we disable the "myths which make inequality appear beneficial, just, or unavoidable"; and third, that we attempt to "unify diverse groups" (1976:249). In considering these three points, they briefly discuss the revolutionary potential of the free school movement for establishing antiauthoritarian, democratic cooperation and then merely mention the potential of such other reforms as greater local control of schools and the use of educational vouchers, a plan they see as having the potential to "equalize educational resources and foster the proliferation of alternative educational settings (p. 262).

Bowles and Gintis do not consider whether or not their separate goals are really compatible. It is one thing to endorse a more equitable allocation of educational resources and the fostering of educational reform; it is a very different matter to assume that there are ready means of attaining these goals. Social reforms would be much easier to attain if one could dismiss categorically as an obstacle to progress the notion that schools are centers of symbolic representation where students learn to negotiate meanings apart from their ordinary context. One could then advocate making schools into microcosms of the larger struggles within society itself.

The problem with such an approach, however, is guaranteeing that the struggles within the school will not simply reflect the struggles between competing groups that already characterize much of social life. Indeed, one might argue that such struggles often provide the very source of many of the myths of inequality that are somehow to be disabled by relocating the struggle from the society at large to the school itself. Such thinking seems to suggest that the school is the source or the primary means of transmission for the racial and sexual biases that permeate our society and is not a center, albeit a limited and often ineffectual one, for combating these prejudices. Would schools really be a fairer place and hence a better model of democratic cooperation if they were to forgo the study of the symbolic world to become an extension of synecdochic practice? While there is racism and sexism in our schools, why should we assume that the struggle to liberate people from such attitudes is better grounded in existing social conflicts than in the symbolic beliefs in justice and equality embodied in the great texts of the past and those yet to be written?

The fundamental problem with Bowles and Gintis's work is its unwillingness to consider the role of literacy or education generally in disabling the myths of inequality and just rewards that so contribute to the whole apparatus of social reproduction. What is the better remedy—the general struggle that already defines social life outside school or the more narrow political struggle that is given focus and direction by the articulation of goals? Do we need to abolish schools as obstacles to social mobility, in other words, or do we need to transform them so that they may be more effective in helping people to recognize and verbalize their aspirations for a better world?

Knowledge is not neutral, as Richard Ohmann makes clear in *English in America*, a sustained ideological attack on the English curriculum and profession, and the ethos of professionalism often supports existing power relations. Educators, whether they like it or not, are instruments in a political struggle. The really difficult question for educators, however, is not how to fulfill their own political agenda but how to enable students to exercise political power themselves. And for English teachers the question is even more direct—how to help students exercise the power that comes from being able to comprehend, articulate, and pursue new modes of being embodied in texts. As literary critic Gerald Graff notes, while the literacy skills that are often taught in writing courses, such as the ability to analyze data and to provide reasons for action, are often useful for corporate expansion (or exploitation), these same skills are also necessary to engage in the "thinking and expression that challenge the technostructure" (1979:106).

It is unquestionably easier for teachers to act directly in pursuit of their own vision of what the world should be like, including creating a world desired by Bowles and Gintis where teachers and students would be equal, but the fundamental, intractable problem is that they cannot create such a world by any direct means, not even within the confines of the classroom, where their power remains embedded in their position despite any efforts to abrogate their authority. Even more to the point, they cannot give constructive assimilative powers to their students, since such powers are ultimately derived either from the physical power to transform the world (which likely neither the student nor the teacher actually possesses to any great degree) or from the theoretical or ludic power to reconstruct the world in symbolic activity.

Indeed, the whole question of the role of education generally in promoting social reproduction becomes much more complex if one considers the possibility that literacy may be valued for the actual power it gives to individuals to transform the world and not just because it is some mechanical school exercise that is more difficult for some students than for others. Raising such a possibility, besides dispelling the revisionist critique that literacy is a form of cultural domination, forces us to confront the one question that is hardly ever asked—namely, what constructive, practical role can literacy itself, not schooling or traditional reading and writing instruction, play in transforming the social world?

III. In revisionist critiques of education, literacy is usually equated with the most visible and most superficial components of reading and writing instruction and hence with those aspects of the curriculum that have arbitrarily been selected to exclude the least advantaged. Literacy is not regularly considered an important, complex historical phenomenon necessarily linked to the material foundation of social life. Although the capacity to use language may be related to certain genetic characteristics of the human brain and although the technology of coding language does rely heavily on arbitrary conventions, literacy itself is grounded in the general human development of symbolic representation and thus reflects the social and historical forces that necessarily affect cognitive growth.

The ability to comprehend and create symbolic verbal meaning, in other words, is valuable not simply because people's levels of attainment reflect their general relationship to production, as might be the case with their geographic distribution, as reflected in their zip codes, or with their access to costly and elective services, as reflected in their dental records; and the allocation of rewards based on literacy is not simply an indirect means of ensuring social reproduction under the guise of fairness. To the contrary, the full development of symbolic representation in language is itself an important component of the heightened accommodative-assimilative responsiveness that lies at the heart of our collective and individual efforts to attain power both over the world and over ourselves. The key to our achieving such power in literacy, as in cognition and social activity generally, is the development of our ability to perform increasingly constructive, deliberative transformations of what is given.

At the same time, however, we must also admit that social power always tends to reside in the transforming power not only of the individual but also of the position that one occupies, and positional power is the more readily transferred across generations. The children of rulers and the children of factory owners tend to be rulers and factory owners themselves, and perhaps to a lesser extent, the children of doctors and lawyers tend to be doctors and lawyers—but with one crucial difference. The power and status associated with such middle-class, professional positions cannot be quite as simply inherited, since they require the certification by educational institutions of the individual's actual symbolic transformative power. The standing enjoyed by professionals such as physicians, lawyers, engineers, architects, and economists in part reflects their considerable transformative powers associated with, if not directly related to, literacy.

It is one thing to object that, despite the intrinsic value of skills in symbolic negotiation, we overcompensate the people who possess them or to make the even more compelling argument that the current system of social reproduction does not give all people equal opportunity to obtain these skills someday; it is another matter entirely to confuse the transformative skills reflecting the heightened development of symbolic representation with style, mannerism, and affectations that are valued largely as accoutrements of power. There is no reason to believe that everything could somehow be readily changed if only everyone could be made to see these skills as reifications.

The value of literacy derives not so much from the prestige of any one group as from the power that actually resides in the general ability to transform the very conditions of our existence, including the power to plan for and promote such changes by objectifying other ways of being in written language. Thus to the extent that literacy is a defining characteristic of a technologically advanced society, then all people, in some crucial sense, must become comparably advanced if they are to play an active role in shaping their own destiny. And here is an irony of recent education reform: while the most visible revisionist critics have been devoting much of their energies to transforming educators and in turn the educational system by fostering a much needed tolerance of cultural diversity (and thus enhancing the status of social groups, which Randall Collins sees as one of the vital historical functions of schools), the task of advocating the transformation of students themselves, especially those from the working class, in order to ensure the development within them of an increased power to objectify and pursue specific political goals has fallen to social critics, with closer ties to traditional Marxism. The Marxist political philosopher Antonio Gramsci, for example, calls for a new form of working-class education, one that will transform the consciousness of workers and their children by freeing them from a consciousness that was largely shaped by the conditions and the constraints of indigenous practice.

Gramsci bases his call for reform on the need to reconcile what he calls the "two theoretical consciousnesses" of the worker—"one which is implicit in his activity and which truly unites him with all his fellow-workers in the practical transformation of reality; and one, superficially explicit or verbal, which he has inherited from the past and uncritically accepted" (quoted in Femia 1981:43). This latter conception of one's world, what Gramsci calls "contradictory consciousness," is immensely important to his whole critique of education. It exists for all of us like Vygotsky's "pseudoconcepts" or Piaget's "figural collections"; that is, it produces thoughts that resemble true concepts but, unlike true concepts, which we freely select as theoretical possibilities in accord with our real interests, these impose themselves upon us with the sense of the necessity and the appropriateness inherent in customary practice.

Our "contradictory consciousness" exists not as the expression of the indigenous practice of the workers themselves and the interests represented by their concrete activity in the world but instead as the expression of the symbolic values of the dominant group—values that often conflict with their legitimate interests. Our consciousness, in other words, is mediated by symbolic meanings that are the product less of our aspirations than of the aims of the dominant social powers. The reversal of identification results in what Gramsci calls the cultural "hegemony" of the dominant class—the dispossessed now judge themselves in the light of the values of those whose practical skills and station in life give them far greater access to power. This "contradictory" or false consciousness, Gramsci writes in a central passage, although "superficially explicit or verbal," nonetheless "influences moral conduct and direction of will, in a manner more or less powerful, but often powerful enough to produce a situation in which the contradictory character of consciousness does not permit of any action, any decision or any choice and produces a condition of moral and political passivity" (quoted in Femia 1981:43).

For Gramsci, therefore, the school fails if it succeeds only in integrating students into the work force, for there will still be a split between those equipped to work and those equipped to rule. While the target of Gramsci's specific attack is vocational education, what he says applies equally well to "functional literacy" and to all efforts of social reform that tend "to perpetuate traditional social differences." By giving everyone a skill, Gramsci argues, vocational education succeeds only in creating the illusion of democracy:

> The labourer can become a skilled worker, for instance, the peasant a surveyor or petty agronomist. But democracy, by definition, cannot mean merely that an unskilled worker can be skilled. It must mean that every "citizen" can "govern" and that society places him, even if only abstractly, in a general condition to achieve this. Political democracy tends towards a coincidence of the rulers and the ruled. (Gramsci 1971:40)

The attainment of true democracy entails not, as Bowles and Gintis suggest, substituting a model of a liberated society for the current pedagogic process but instead strengthening that process, even at the cost of making it less democratic, so that it might be better able to produce students whose powers of symbolic representation in language and elsewhere are the equal of the teachers themselves. Gramsci clearly realizes, as American radicalism for the most part does not, that, if society itself is ever to be changed by people working collectively in their own interest, then students themselves, more than their teachers, need to be transformed. To the extent that literacy has a vital role to play in self-transformation, all students thus need to become fully literate readers and writers of texts, and this attainment must be the primary goal of education.

Universal literacy, however, cannot be attained in a vacuum. At times the general direction of historical change may be acting as an impetus for increased literacy. Bernstein notes, for example, that the political struggles of American blacks were a natural catalyst for more symbolic language use; since the whole purpose of the civil rights movement was to create new opportunities for action (in other words, to widen the scope of accommodation, thus making it more deliberate), it is hardly surprising that it should simultaneously have generated more constructive assimilative responses as blacks became aware of a new range of choices and thus began to make more elaborate plans of action. Implicit in Bernstein's writings, therefore, is a plan for enhancing the literacy and hence the standing of any dispossessed group.

Bernstein's research, however, has come under continuous attack from revisionist critics for two decades now, in large measure because it ignores the question of the speech of the dispossessed and its status, seeming indeed to support prevailing attitudes by devoting so much attention to analyzing and thus seeming to justify the role of language in existing patterns of social reproduction. Bernstein's concern with the social basis of symbolic representation in language has been dismissed by the prominent thinker Noam Chomsky as "reactionary in its implications and perhaps hardly worth discussing as a specimen of the rational study of language" (1979:56). In such an ideologically charged world, the concern with improving the literacy and the concern with improving the status of the dispossessed, although both laudable goals, can well be at cross-purposes.

For the most part, few social critics have been willing directly to address the question of how economic and cultural deprivation affect the development of literacy and symbolic representation generally. Whereas educators, like ethnographers, avoid the seemingly pejorative term "deprivation," British educator Harold Entwistle notes that the very idea of deprivation is "built into the notions of exploitation and alienation," without which it becomes difficult to "justify the disturbance of existing ways of life which is implicit in even modest social democratic proposals for reform" (1978:86).

British Marxist Maurice Levitas is even more outspoken in rejecting the plea that the principal task in social reform is to make the existing social order more accommodating to the conditions in which people happen to find themselves living: "It is unnecessary," he writes, "to search for formulae intended to hide the poverty of the working class, or to label that poverty as being merely a quality of a different social relation which requires a different learning process which in turn needs a different linguistic code" (1974:148). For Levitas, there is an indisputable advantage in widening one's perspective so that verbal meanings acquire a universal rather than a particular significance, and the educational problems entailed in attaining such a perspective cannot be divorced from the economic basis of social life. The capitalist, he adds, stands "in relation to production as witness to its entire process and in relation to the market as witness of the economic process as a whole. Thus, his economic status has as its complement in roles whose meanings have a greater universality than have his clerks and his overseers, his managers and his scientists" (p. 147).

Starting from a position of economic, cultural, and verbal disadvantage, therefore, the working class or those segments of the general population with a simpler, more immediate connection to production must transform themselves in the very process of transforming society. Education, in other words, always entails a dynamic process of self-transformation both for individuals and for groups. People must change themselves in the process of changing the world, or, as George Lukács says in his major work on social theory, people's struggle against their conditions must include a struggle against the attitudes and even the consciousness that is an integral part of that condition:

Thus we must never overlook the distance that separates the consciousness of even the most revolutionary worker from the authentic class consciousness of the proletariat. . . . *The proletariat only perfects itself by annihilating and transcending itself, by creating the classless society through the successful conclusion of its own class struggle.* The struggle for this society . . . is not just a battle waged against an external enemy, the bourgeoisie. It is equally the struggle of the proletariat *against itself:* against the devastating and degrading effects of the capitalist system upon its class consciousness. The proletariat will only have won the real victory when it has overcome these effects within itself. . . . The proletariat must not shy away from self-criticism, for victory can only be gained by the truth and self-criticism must, therefore, be its natural element. (1971:80–81)

Lukács's tone and vocabulary may make his central point seem somewhat foreign and unrelated to contemporary social problems, but this point, that those who wish to transform the world must also transform themselves in the process, is still an important response to the urgings of reformers for some sort of collective social conversion. Unlike the characteristic plea of American reformers for tolerance and cultural diversity, Lukács's plea is for social reform and political equality. While education should make students more aware of cultural differences (as part of the very process of deepening symbolic representation), it cannot have as its goal the maintenance of this diversity, as if groups of people are to be preserved as endangered species. Each group, like each individual, must be made conscious of its historical situation and of the way in which that situation can be altered through self-effort. E. P. Thompson, in his classic study *The Making of the English Working Class*, describes the regular and at times heroic efforts of workers who struggled to give voice and power to their collective aspirations. Their story is one of self and social transformation and includes as one important ingredient efforts "to raise the level of political awareness" (1963:717) by the formation of reading and debating societies.

Teachers then and now have an important role to play in the whole process. At the very least, they need to transform themselves by continually making themselves, other educators, and the public at large more aware of the needs of their students. In this regard, revisionist critics of education and literacy have led the way in promoting Marx's third thesis on Feuerbach, "that circumstances are changed by men and that the educator himself must be educated" (1967b:440–41). Even more important, however, educators have the task of educating students, particularly of making them aware of the contingent nature of the entire social world, including both that which they want to enter and that in which they were raised.

Students who have as their highest goal the preservation of their cultural identity will likely have to work harder, learn more, and confront more difficulties than those who are willing to forsake their past, just as it is now necessary for those who want to preserve the natural world to gain a greater understanding of technology than those who wish merely to continue its commercial exploitation. The essential difference, however, is that, except in unusual cases of debility, people should not require custodial protection; they should instead be encouraged to act in their own interests, even if there is an inevitable conflict between an individual's or a group's existence at any one moment and the emergence of those skills that allow people to transcend the boundaries of the present. Ultimately, teachers of literacy must speak for what can be, as compared to what is—even if that future is only to be a preserved and protected form of the present.

Meanwhile, the students themselves must be led away from the present, not as an act of rejection, but as a means of gaining the objectivity that comes from seeing the contingent nature of reality. The understanding that we seek in education and in literacy is always an understanding conditioned by our living in a world whose meaning, like that of a text, must be seen as something apart from ourselves, that is, as a theoretical possibility, if it is to be understood as a human and not a natural phenomenon. In this sense, a typical freshman writing assignment such as "How Are You Different from Your Parents?" drives at the heart of literacy in asking students to objectify an understanding of their world as a means of recognizing the difference between what has existed and what can be. Students do need to feel accepted, not defensive, where their past is concerned if they are ever to deal with it objectively.

In order to facilitate growth, the teacher must therefore always stand in two worlds—the one from which students have come and the one to which they are going. Moreover, the teacher's acceptance of the students' past cannot be a duplicitous act aimed at tricking people to turn against their past; instead, it must be an act of true sympathetic understanding, grounded upon the recognition that symbolic growth proceeds by rejecting the present in the imaginative act of reconstructing it. British educator Richard Hoggart describes his own ordeal as a "scholarship boy" moving from a primarily synecdochic world to a primarily symbolic one in the final chapter of *The Uses of Literacy* (1961), a work that derives its considerable literate powers by imaginatively reconstructing English working-class life. In a more recent essay, Hoggart warns against the danger of romanticizing those with different life-styles, adding that our obligation is "to help them, whilst not doing wrong to whatever may be good in their present worlds, to help them in the right ways, to—and I choose the verb deliberately— *surmount* that world" (1982:85).

The whole process of educational reform cannot take place outside the very historical conditions that gave rise to and continue to support current practices, just as those responsible for implementing significant educational reforms are not just instruments of change but part of the very historical process in need of transformation. Literacy education especially is itself necessarily conditioned by the very historical forces that first produced and still continue to reproduce gross inequities in the access to literacy. The process of socialization does produce groups of students who, relative to other groups, are less well prepared for the deliberate, constructive symbolic representation that lies at the heart of literacy. This inescapable historical fact impinges upon all discussion of reforming literacy education, revealing at almost every turn reasons why true educational reform is so easy to advocate and so difficult to achieve.

Two separate factors, however, mitigate this difficulty somewhat, allowing for significant educational and social change, for better or worse, independent of whatever success we may have in improving literacy education. First, the enhancement of status has always been and continues to be a significant function of schooling, and second, the structural connection between literacy and social advantage is as much theoretical as practical (that is, a great deal of wealth, power, and prestige is transmitted directly between generations without the significant intervention of any tests of symbolic representation). There is thus the possibility that the actual importance of literacy or symbolic representation generally may be weakening despite current talk about the generally higher cognitive demands of our age. There is no reason, in other words, for the connection between literacy and social structure to remain unchanged, especially if our concept of literacy and our system for allocating rewards should radically alter. It is quite possible, for example, for a society to reward the interpersonal communication skills vital to being a good salesman far more than those required to be a good writer. Given the cultural and psychological importance of literacy, however, it is questionable whether such changes in the social order that result in the diminution of literacy's role in social reproduction will necessarily be beneficial. As imperfect as the current system for allocating rewards is, we need to resist the assumption that any changes to it, including diminution of literacy's value, are necessarily an improvement.

In any case, educators certainly do have the option of working to undermine the system that rewards the level of symbolic representation indicative of literacy, either by promoting the status-conferring nature of schooling or by urging that we diminish the symbolic component of the school curriculum generally, including, of course, work to diminish the attention paid to the reading and writing of texts. The one thing that educators cannot do, although many would like to, is to promote *literacy* in conjunction with these other two goals—unless of course "literate" is redefined so that the symbolic component is diminished. Such a tactic, however, can have only limited effect as long as the level of symbolic representation necessary for problematic literacy continues to play an essential role in determining the pattern of social, political, and economic reproduction. In such a world, the historical condition that makes the attainment of literacy ordinarily more difficult for poorer students than for richer ones remains a necessary part of any solution short of the radical reorganization of society characteristic of only the most extreme and usually the most violent political revolutions.

Educators can therefore proceed only in recognition of the contradiction that underlies their position—namely, that the students who have the most to gain from what is taught in school are by definition the least equipped to profit from the opportunity. Indeed, it may even be possible to define the governing principle of middle-class family life as that which enables children to succeed in the symbolic negotiation that ideally forms the center of school curriculum. Such a recognition, however, is hardly reason for despair or for turning away from literacy education. Instead it calls for an increased commitment by all to overcome initial disadvantage in achieving the goal of equality of performance; we should not be content with our efforts in education until there are no vestiges of social origins in the measures of the students' literacy. Even more important, such recognition also indicates the direction in which we should be moving to improve literacy education: we should attempt to establish within the school the structures for learning for those whose socialization gives them an initial advantage. Specifically, we need to worry less about the content of what we teach and more about the complex structure of motives that compel students to see any specific lesson, or their entire education, as serving some long-range developmental need.

Universal literacy education, if it is to be effective, must ultimately be based upon students' use of symbolic representation for personal goals. Reading and writing must become practical activities that develop direct connections between, on the one hand, the necessarily theoretical experience of reading and writing texts and, on the other hand, the arousal and fulfillment of the student's own practical goals. Thus it is wrong to speak of literacy education, in E. D. Hirsch's terms, as consisting of "specific contents as well as formal skills" (1983:162), which suggests that what must be nurtured is either abstract skills or specific knowledge of certain important texts, when the real issue is nurturing within students a critical, imaginative, creative attitude about their condition in the world. Hirsch is rightly concerned about the numbers of students who "cannot understand newspapers" because they lack this shared knowledge, but in contrasting skill and content he reflects, rather than answers, the problem at hand. Indeed, the reading of texts in order to ensure that students have an "appropriate, tacitly shared background knowledge" is apt to be as formulistic and hence as removed from the true cultural conditions of students as the current language instruction he so deplores.

The literacy education Hirsch advocates, drawing upon the dichotomy of skills and content, is likely to alternate between being purely practical (as is the case when the goal is learning "survival skills," for example how to complete a job application) and purely theoretical (as is the case when students studying Milton are unable to make any relevant connection between what they are reading and writing and their own historical predicament). It will not provide, however, what is most needed—namely, a literacy education that seeks constantly to reconcile the extremes of theory and practice, drawing upon what one critic of Lukács describes as the "need for the practical concern of *conscious* mediations, to realize in practice at the level of concrete totality the historical tendency whose mediations have become conscious in theory" (Kilminster 1979:83). That is, educators most need ways of confronting students both as readers and as writers with texts that will objectify for them the insight into their historical condition, which otherwise exists for them as intuition, if it exists at all.

Lukács is interested in the importance of praxis generally, not the branch of praxis that is literacy, but the connection between praxis and literacy is nevertheless compelling, since no other practical activity is as valuable in uniting the possible with the real. What we read and write above all else must become a "*conscious* mediation," a deliberate effort to objectify our experience of the world. According to Kilminster, Lukács's program for social reform calls for a "struggle against the suppression of mediations through 'practical-critical' activity which dissolves immediacy" (1979:83) and in so doing reveals the contingent nature of what until now has seemed all too solid. Lukács's program for social reform, in other words, is essentially a plea for universal literacy.

What is at stake in organizing a curriculum that will make literacy a form of praxis is much more easily stated than achieved. Within the classroom, the goal must be less to persuade students of the importance of literacy and more for them actually to read and write texts in the very act of clarifying and pursuing goals. "Teaching should be organized," wrote Vygotsky over a half century ago, "in such a way that reading and writing are necessary for something. If they are used only to write official greetings to the staff or whatever the teacher thinks up (and clearly suggests to them), then the exercise will be purely mechanical and may soon bore the child; his activity will not be manifest in his writing and his budding personality will not grow" (1978:117).

There are many students who very much want to become highly literate, just as there are many teachers trying earnestly to help them, but what frustrates them both is the actual difficulty, even given best efforts, of achieving Vygotsky's goal of creating real motives for students to read and write within the necessarily artificial context of the school. Here the role of literature can play a crucial role in creating imaginative situations that nevertheless arouse real motives, just as we can now see more clearly the importance of considering rhetorical factors such as audience in teaching people how to write.

In addition the work of Bernstein and Bourdieu is most helpful in pointing out the complex, historical nature of the teacher's dilemma in trying to make what transpires within the confines of the symbolic classroom equally relevant to students from diverse backgrounds. Students, parents, and even teachers unfamiliar with the symbolic demands of literacy often find themselves eager to compensate for their deficiencies by making a more strenuous application of the very synecdochic practices like handwriting, spelling, and punctuation that constitute the machinery of literacy, while the essence of literacy itself, a psychological propensity to objectify verbal meanings not shared by the group, goes unnourished.

As Bernstein observes, there is no ready means of overcoming historically entrenched disadvantages. He notes that the alternative of the open classroom, for example, poses almost as many problems as the traditional means of class governance, at least in the early years of schooling, since, for those students ill equipped to deal with symbolic meaning, the absence of a clearly defined synecdochic context is apt to be especially disorienting, while the children from professional families are already familiar with the task of seeking out and following the implicit structures that really govern the class. Bernstein's analysis leads him to a realistic assessment about the prospects for educational and social reform. First, while never optimistic, he insists that the "causal relationship between the structure of social relationships and the structure of communication" (1977:30) does offer solid grounds for directing positive change. We can study just how literacy develops or fails to develop within various groups and with this knowledge can begin working to change existing patterns of reproduction.

The reforms themselves, however, can be implemented only in a world characterized by the permanence and the obdurance of the very social relationships that regulate the existing pattern of social repro-duction, thus mitigating many of the effects of any reform, as the groups originally in an advantageous position continue to respond to the opportunities offered by any reforms (and likely organize opposition to the others). Thus there is Bernstein's second, more pessimistic point—namely, that it is foolish to deny the "pervasiveness of the stability of power relationships and the many forms of their transfor-mation," obviously including the many educational and social reforms that, in promising more than they can ever deliver, seem too often to end up legitimatizing rather than transforming existing patterns of reproduction.

IV.   Schooling, literacy, and social reproduction are inevitably linked in a circular pattern of cause and effect, and as with the hermeneutic circle itself, what is most important is not to destroy the circle and thus to distort reality but to enter it in such a way as to foster under-standing. The family and the school, as Bernstein states, are not them-selves "major levers of radical change" (1977:xx), and thus we should realistically expect educational reforms to reflect significant social changes rather than to precipitate them. People concerned principally with altering the pattern of social reproduction, therefore, are doomed to frustration when they devote their full efforts to education and would probably do better to work directly in the political arena. More-over, to the extent that teachers confuse the educational and political arenas and thus identify social changes within the classroom as their major goal rather than psychological change within their students, they negate the limited but still vital liberating potential of literacy. Teachers must be less concerned either with acting directly for their students outside the classroom or with creating the illusion of equality within the classroom and instead must be more concerned with ensuring the freedom for their students that comes with the deepening symbolic powers associated with the mastery of literacy.

Political equality characteristic of society outside the classroom is an uncompelling model for the ideal equality between student and teacher, since, as noted at the end of chapter 5, democratic consensus does not guarantee freedom or the rights of minorities or even democracy itself as much as does the persistence of the group's allegiance to symbolic meanings as constituting ideals whose existence transcends synecdochic practice. It is a mistake, therefore, to confuse any one governing principle of the existing political practice—even, for example, that the majority rules or that the individual is entitled to an opinion—with the more essential, overriding commitment of the group to the belief that individuals should be encouraged to pursue truth not shared by the group. By placing this commitment to symbolic truth at the center of reading and writing instruction, the teacher works to guarantee the integrity of the political system outside the classroom; to replace the symbolic freedom of the classroom with a miniature version of the political system itself is to confuse the imperfect machinery of democracy with the true spirit that guarantees its survival.

Teachers of literacy cannot really be concerned with a specific social agenda; it is not enough to substitute one set of synecdochic meanings for another. While literacy, in opposing meanings based upon existing social relations, is always a liberating force, it is nonetheless unfair to equate symbolic meaning generally with liberal reforms and synecdochic practice with a specific conservative ideology. Conservatives as well as liberals see themselves as opposing current practice and in their opposition also call for change based upon symbolic principles. Moreover, as we shall see more fully in chapter 7, the maintenance of certain traditions may best ensure the continuation of literacy, while specific reforms, despite their popular appeal, represent new and more powerful forms of synecdochic domination.

Not change itself but the spirit of renewal and reform, so important for literacy, guarantees the continuation of freedom and of democratic institutions—the liberating power of literacy comes only in the recognition of the contingent nature of social institutions, not in their necessary abolition. To see the educational process and students themselves as instruments of the teacher's own desire for symbolic transformation is itself a fundamental violation of the spirit of literacy, however effective it may be as a tactic for social reform. Moreover, we mistake the nature of literacy itself when we fail to consider how the benefits of any social reforms may affect the continuity of the literate tradition itself. Certainly some conditions are more favorable than others for the furtherance of literacy, and as citizens we should work to make such conditions universal. As educators, however, we can work for this political agenda only indirectly, either by direct political action outside of the classroom (in which case we are not acting directly as teachers) or indirectly within the classroom by giving the students the power to decide for themselves (in which case we are not acting directly in support of any program).

As educators, therefore, we serve the future by serving literacy, and in turn we serve literacy by judging students not on their ability to give correct answers but on their ability to comprehend and create texts embodying new modes of being. Our first priority, consequently, has to be the continual effort to make the pedagogic process itself increasingly more symbolic; we must work to ensure that it is better able to overcome the inherent obstacles posed by the existence of class and by patterns of social reproduction generally. While the educational establishment must assist any group in its effort to objectify its existence through the reading and writing of texts, it also must resist the constant pressures merely to reward the status of entrenched or ascending social groups, for to do otherwise is to further the school's role as an arena of political struggle.

Radical critics are quick to note that schools are already arenas of political struggle, and no doubt they are, but much of their legitimacy comes from their position as centers for symbolic exchange. It is mistaken to see the great influence of schools in all technologically advanced societies as due to their inherent power and not to the pervasive acceptance of the importance of the concept of symbolic representation. While much of the symbolic knowledge conveyed by the schools, especially in the sciences, is inherently powerful and crucial for the continuation of the existing social structure, it is far from certain that a society would have to establish and maintain a system of universal literacy education in order to maintain and even extend its technological control over nature when the training of a technocratic elite might suffice. In addition, a government that wanted to extend such technological control over its own people might well be better off letting literacy education more accurately reflect existing social discourse, with its peculiar mixture of partisan haggling and widespread apathy.

Only because schools are so widely accepted as centers for symbolic negotiation are they such easy targets for those with specific political agendas, from both the right and the left. It is not especially difficult for a group to gain control of a school; it is difficult, however, to maintain the school's legitimacy in the public eye when it is apparent that it is not principally concerned with symbolic exchange. Political groups from both the left and the right are mistaken in their belief that they can maintain the legitimacy and hence the power of schools while making them the instrument of synecdochic practice and belief. They are mistaken in believing that the schools themselves have power that is independent of the pervasive acceptance of the role they have in social reproduction, either by conferring status on groups whose base of power is well established outside of schools (the more traditional function of schools) or by certifying the mastery of synecdochic and symbolic practices (the more modern function of schools). It is naive, in other words, to believe that the theoretical potential of schooling to break the traditional mode of social reproduction is inherent in the institutions themselves and not in the broad-based public faith in the value of symbolic transformation.

Critics who accuse schools of maintaining traditional patterns of reproduction condemn the present system as a failure compared with an ideal arrangement that has never existed and, given the historical connection between symbolic representation and class, can perhaps never be fully attained. In part such criticism is itself the very model of literacy, relying as it does upon an ideal sense of what should exist. An understanding of what already exists in the present or has existed in the past, however, depends equally upon the imaginative power of literacy to reveal what, at least to others, is not there. And here radical critics often fail to perceive the current system of public schooling for what it is—a limited, imperfect improvement on past practices of reproduction that is constantly being threatened from two quarters.

One danger is that schools may come to embody more completely the divisions, prejudices, and passions already embedded in society, becoming in the process even greater instruments of social reproduction than they are today. If they did so, the large corporate powers and government agencies that still need people with narrow technological training would then likely turn away from public schools governed by public policies that have the intent, if not the power, to provide more equal opportunity and establish their own training centers—with literacy education and the ideals that it embodies left to languish, unsupported by either a politicized public educational system or a technological private one.

The other danger is that society as a whole can come to have less use for a literate population, less use for the ideal that people generally need to be able to read and write texts embodying anomalous ways of being in the world. The prevailing institutions that run society will undoubtedly continue to have a great need for a corps of technologists and bureaucrats capable of maintaining the system, but we should not assume that this elite cadre, much less the population as a whole, will have to be literate. After all, it is quite possible for the masses of people to become increasingly removed from the technological, fiscal, and political powers that control their lives, and if they are, the schools are likely to reflect the passivity, the parochialism, the political conformity, the dependency, and the division into unconnected specialties that are already coming to characterize the workplace. It is reckless and short-sighted to see such a condition as necessarily fostering more progressive forms of social reproduction or even radical change. Such a condition is just as likely to foster a world of political apathy and conspicuous consumption as people, increasingly unable to see the contingent nature of the powers that control their destinies, turn more and more inward.

The proper role of the school in the face of such a prospect is to promote literacy as a "practical-critical" activity, to help students to create meaningful verbal structures that are independent of the dis-orienting, synecdochic pull of social life and that in turn promote within students the ability to recognize the theoretical and hence contingent nature of the social world. It is Gramsci's compelling insight that "common sense," or the unreflective understanding of the people, is less the refuge of collective wisdom and thus less the protection against totalitarianism of all sorts than the limit imposed by the dominant social group as to the range of phenomena that either can be changed or must be accepted as given (1971:325–43).

"Good sense," for Gramsci, however, is something quite different: it represents the questioning spirit that seeks to move us to act in our own interests by arousing within us an awareness of the possibility for change, and as such it is closely aligned with the self-critical spirit of philosophy itself. Education then consists in nurturing the "good sense" of people to think and to act for themselves; yet as Gramsci warns, such a goal cannot be achieved without a dialogue between the present, as embodied in the group itself, and the future, as embodied in those who can articulate other ways of being. There is no progress, in other words, without a literate elite committed to the education of the community.

Schools and educators play a dual role in this process: first, by being part of the elite and thus helping to educate the community directly, and second, by ensuring the existence of leaders able to articulate a vision of the future and thus helping to educate the community indi-rectly. Although the direct approach may seem more efficient, it is doubtful that it can be sustained without the continual production of new leaders with necessarily closer synecdochic ties to their generation and their community itself than the educators who trained them. Thus while the long-term goal of schools may be the transformation of society, the immediate goal has to be the transformation of students, and this transformation is best achieved by pedagogic practice that leads to the highest standards of literacy.

Bourdieu and American social critics are both correct in noting that the current emphasis on academic attainment within education does generally result in the reproduction of the existing social order. They err in failing to recognize that schools as presently constituted do little more than measure the levels of literacy and symbolic representation as a whole that students attain as part of their overall socialization. We lack evidence that schools themselves are at all effective in teaching literacy. Indeed, it is doubtful that, except in terms of educational theory and models of curriculum reform, instruction in literacy has ever played a great role in language education. It is less doubtful that the revolutionary, constructive potential of education exists only in the commitment to helping all students, regardless of background, to become fully literate. While certain systems of social organization are more likely than others to foster such a commitment, the unavoidable irony of literacy education is that there is no simple and direct way of creating such a world. There is no means of reforming society so as to ensure that people would act for themselves in the future apart from encouraging them to act for themselves in the present. Thus, however much we may condemn current educational practice, we cannot escape the need to reform, and to continue reforming, the existing system of literacy education.

7

Conclusion: The Future
of Literacy

I.   Literacy's future is inevitably linked with its past. Only the smallest percentage of people in all of history have ever learned to write, and even today outside the industrially advanced countries many people remain illiterate or literate only in the unproblematic sense; that is, they can encode and decode speech but are at a loss when confronted with the task of reading or writing texts. The very idea that being literate entails the ability to comprehend and create texts, although now widely taken for granted in some areas of language education, has had wide currency for barely a century in the United States. While there may be a general support for literacy education today, it is never entirely certain just what is being supported. Over the last twenty years, the world "literacy" has unquestionably had lofty connotations, but its detonation seems to have been increasingly broadened to refer to the wide range of skills, verbal or otherwise, that enable people to do what they want or, in some cases, to do what others want of them.

Moreover, a number of contradictory motives seem to underlie this popular concern with literacy. For instance, we want people to have the skills and attitudes that will enable them to provide for themselves and to adjust to existing social conditions. To this extent, we do want people to be able to produce the "correct" response when they read and write. We also sense the historical connection between literacy and social standing and thus intuitively recognize that the promotion of literacy is a form of class hegemony, both allowing students with the "right" backgrounds to succeed and encouraging those with different backgrounds to develop skills and attitudes that will better enable them to compete. In addition we continue to recognize an ill-defined but nonetheless essential value in being able to read the great texts of the past and to write our own, albeit lesser, texts today.

Amid such conflicting motives, the future of literacy in one limited sense seems assured: there is little reason to doubt our ability to produce workers capable of maintaining and even improving the existing economic system; therefore, as long as we continue to use "literacy" to

refer to purely *functional* skills, we can likely expect a continual increase in the nominal rate of "literacy," or at the very least, we have no
reason to expect its precipitous decline. The implications of this sce-
nario for people's ability to read and write symbolic texts, however, are
another matter entirely. Indeed, one problem we face is that the rates
of "literacy" may appear to be growing at the very time when the actual
ability of students and the population at large to read and write symbolic texts is in decline, and perhaps most troubling of all, the two
trends may be related.

Behind this concern lies the recognition that literacy is, in Ricoeur's
words, an extension of the general power of metaphor "to break
through previous categorization and to establish new logical boundaries
on the ruins of preceding ones" (1978:131). Literacy, as such, is discourse characterized by symbolic reference to new ways of being,
constituted in texts themselves rather than in deictic reference to what
is already embedded in social practice. Thus, as part of the process of
symbolic representation that Ricoeur calls "the dynamics of thought
which breaks through previous categorization," literacy fulfills a contradictory social role. On the one hand, it is directly related to the
creative, anomalous thinking responsible for creating the world we live
in; on the other hand, it works constantly to destroy or to reestablish
what presently exists and hence what is necessarily supported by a
constituency with legitimate reason to be suspicious of change. Some
symbolic meanings, like the very idea of "literacy" as it emerged over
the last century, do eventually become synecdochic as they are incorporated into the social order or, in the case of literacy, into pedagogic
practice. For such changes to take place, however, the existing social
order itself must change, and since it is always far easier to manipulate
the social order theoretically than practically, symbolic and synecdochic
communication inevitably conflict.

The situation in which we as humans find ourselves at any one time
is both the product of an earlier symbolic action and a contextual
constraint against which we have to act in order to grow. Often we do
not recognize the contingent nature of our historical situation; we do
not recognize the extent to which it was created in response to certain
conditions and can be altered as the situation itself changes. Instead
our historical situation often seems obligatory, as if it were a natural
force before which we must accommodate ourselves in order to survive.
Because we lack a sense of power to alter circumstances, our assimilations remain fanciful and playful, more characteristic of the magical
transformation of children than the constructive transformations of
great artists. At the same time, we do not sense that our inability to
change the world constructively greatly enhances the accommodative
pressures on us to act in accord with custom.

A virtue of literacy is that it offers us a powerful means of breaking the tyranny of the present by deepening the pattern of symbolic representation and thus allowing us to deal with the highly constructive yet theoretical assimilations that form the content of all literate texts. We must recognize, however, that the opportunity offered by literacy is not the necessary consequence of the availability of a mechanism for phonetic transcription. People who learn how to encode do not automatically use that skill to transform ordinary language into the complex verbal assimilations of texts, nor do people who learn how to decode language automatically accommodate themselves to the complex verbal creations of others. Literacy, in other words, does not develop apart from a more general pattern of symbolic representation by which people seek to exercise greater control of their own lives by acting in accord with what Kenneth Burke calls our ability to "interpret our interpretations" (1984:6). The single greatest obstacle to the development of literacy, therefore, is people's lack of psychological motivation to form ever-deepening symbolic representations. Literacy seems to flourish only in the rarest historical circumstances, when, from an active interest in shaping their own future, people are encouraged not just to recite and copy down what is already known but to read and write for themselves.

Such historical circumstances include periods of rapid political and technological change, when new, symbolic modes of being clearly exist as imminent possibilities, not as fanciful speculation or revolutionary dogma. Historically speaking, to ensure the promotion of literacy a situation seems to be needed in which parents immediately sense the advantages of fostering creative, anomalous responses in their children. This situation is perhaps not as likely to occur in the absence of a system capable of encoding speech, and thus it may be correct to say that the change charactertistic of modern life is made possible by the emergence of a system of phonetic transcription; it is incorrect, however, as a number of theorists on literacy have suggested, to see the systems of phonetic transcription themselves as the cause of these changes, since in group after group the technology for coding language has lain fallow as traditional patterns of cultural reproduction continue essentially unaltered. Neither the technology for the mechanical storage of language nor the need for the greater accuracy that such storage supplies as compared with memory seems to provide the basic motivation for literacy. Instead literacy seems propelled forward by the pervasive social need for anomalous responses both to cope with changes that are already occurring or are imminent and to plan for new ones made possible, or at least imaginable, by the general deepening in the individual's power for symbolic representation.

In this light, it is difficult to imagine the pedagogy that could support universal literacy in the absence of a widespread sense of individual or collective historical progress, just as it seems equally difficult to supress literacy through direct political and pedagogic action when people have both access to some technology for writing and an active interest in improving their condition. It is likely as difficult, for example, to promote literacy in a country like Haiti as it is to suppress it in a country like Poland; more important, in neither case would much be accomplished by discussing the future of literacy strictly in terms of pedagogic or curriculum reforms, apart from questions of widespread social and political changes.

Similarly, the future of literacy in the United States and in other countries at comparable stages of economic development is finally a question concerned less with educational practice and more with the evolution of culture itself: what kind of world are we in the process of creating for ourselves? The answer we reach may be less a practical solution or guide to the future than a historical analysis of our condition. Specific pedagogic and political reforms may be our goal in studying the question of the future of literacy, but we can only begin with an accurate understanding of the factors currently at work both supporting and undermining the historical basis of literacy itself. We may not be able to change any of these conditions readily, but without the historical understanding of them, we can hardly know what to support and what to oppose when contending factions clash, each proclaiming itself to be the champion of literacy.

II. Literacy is the inner speech of the community—in written language we constantly judge our present perceptions of the world against a verbal objectification of other possibilities of being. In reading and writing we construct what is not already present as a means of generating such possibilities. The content of literacy, in other words, always exists as theoretical knowledge, as that which is present only as something *out there*, not already grasped by the reader and the writer on the basis of their previous social experiences. What threatens literacy, consequently, must also threaten the possibility of theoretical knowledge itself, at least as such knowledge plays an essential role in people's practical activity. Conversely, factors such as political censorship and new technologies for communication are threats not in themselves but only to the extent that they diminish the practical value of theoretical activity and in turn our ability to engage in verbal forms of such activity.

As already noted, there is good reason to believe that powerlessness, embedded in the technology of a group and manifested in the individual's lack of a sense of self-importance, has been the single greatest obstacle to literacy historically. It almost seems as if our collective ability to transform the natural and social world must change visibly within each generation in order to promote the proper climate for literacy. One might thus assume that technology itself, as the power to effect such changes, is one of the main sources of literacy and hence that the continual growth of our technological mastery of nature is the surest guarantee of the future of literacy.

Yet such an approach is mistaken in equating the actual transformation of nature with the active sense within individuals of a direct connection between their practical and theoretical activities, that is, between what they desire and what they can achieve through their own efforts. Our technical ability to transform and now to destroy the natural world may well continue to grow, but what is crucial for the survival of literacy is not the maintenance of this power itself but the maintenance of the sense of tension within people between what is and what can be, for this very tension between stability and change spurs creative, self-directed activity. In the language of cognitive development, people must have the motivation to attempt more constructive assimilations so that in turn they have the freedom to opt for more deliberative accommodations.

Social groups with minimal powers to transform nature are more likely to see any such power as magical while at the same time demanding strict imitative practice in many areas. The combination of strict accommodative practice and a belief in magical assimilations thus works together to disguise the contingent nature of the group's historical condition and in so doing effectively reduces the motivation toward praxis. The threat to literacy that we face today is not all that different, although ironically it comes from our being in the opposite position of having what seem to be maximum constructive assimilative powers. The danger here is that our assimilations will also *seem* magical, not because they are, but because they are so immense, so thorough, and, most important, so inadequately understood by most of us, controlled as they often are by a technological elite. Once we believe that the world can be changed almost magically, there is less need to adjust to it and less need for any accommodation whatsoever, and without a constant sense of the need for accommodation, there is no pressure to deepen the representational equilibrium that lies at the foundation of all human symbolic activity.

Just as the lack of constructive assimilative power leads to the prevalence of strict accommodative practice, so the total dominance of assimilation may lead to the collapse of most accommodative pressure. The historical development of the assimilative power of technology thus moves us in two different directions—toward the increasing technological control and finally domination of nature and toward the increasing array of life-styles and models of cultural expression. As our power over nature increases, the realm of practical-theoretical activity, where we struggle as individuals to make the possible real, changing ourselves in the process, seems to be in constant danger of diminishing.

On the one hand, there is the allure of experts for whom our desires represent not a personal struggle or an act of praxis but the consummation of the accommodation that defines their professionalism; on the other hand, there is the allure of alternative forms of accommodation that have in common both that they are the manipulated creations of other professionals and that their common theme is the ultimate cultural value of adaptability. Collectively we celebrate both professionals who provide directly for our needs and the personal freedom that allows us to seek life-styles that often prove to be little more than the manipulated products of other professionals. As the French philosopher Jacques Ellul writes, "Advertising, mass media entertainment, political propaganda, human and public relations—all these things, with superficial divergences, have one single function: to adapt man to technology; to furnish him with psychological satisfactions, motivations that will allow him to live and work efficiently in this universe" (1980:313).

The technological mastery of nature, therefore, has within it the force both to inspire literacy in the form of acts of praxis and to destroy the very tension that makes possible all praxis, including literacy itself. The German philosopher Hans-Georg Gadamer, like Ellul, sees the greatest danger to the common civic life on which literacy must be based as the "ideal of a technocratic society, in which one has recourse to the expert and looks to him for the discharging of the practical, political, and economic decisions one needs to make" (1981:72).

There are two basic problems here. First, the actions of the expert, including those of the literary critic, the apparent embodiment of problematic literacy, are often nothing more than a heightened form of synecdochic practice—knowing the correct way to plug variables into formulas while shrouding the whole practice in a specialized vocabulary. Second, the existence of an elite corps of experts relieves the rest of us of the need to struggle in order to address our own problems.

Rather than having the masses of the population engage in praxis, we need only a small professional class catering to and creating our needs. As Gadamer writes, "In a technological civilization it is inevitable in the long run that the adaptive power of the individual is rewarded more than his creative power" (1981:73–74). Only a few are needed to plan and manage, while the majority are free to consume, and as German social theorist Jurgen Habermas concludes, echoing the terminology of anthropologist Robin Horton, any civilization "devoid of the interconnection between theory and practice . . . is threatened by the splitting of its consciousness, and by the splitting of human beings into two classes—the social engineers and the inmates of closed institutions" (1983:282).

If literacy is to have a future, this vital connection between theory and practice—the arena of praxis or practical-critical activity in which individuals are encouraged to create new forms of being—must be preserved. The importance of the constructive assimilative element of literacy is so apparent at this point, since reading and writing are inherently acts of praxis readily available to all. Similarly, we can best see here the threat of technological innovations in mass communication, which are capable of making re-creations of events increasingly life-like, possibly reaching the point where synecdochic experience itself, and not some imagined representation of it, seems to be the content of communication. The danger lies in the possibility that such sensory-rich communication may become an extension of the general assimilative power of the culture as a whole. Like a technologically dominated world, sensory forms of communication threaten to relieve us of the task of having to change ourselves in the process of interacting with them. In reading and writing and even in a medium such as radio, the constructive assimilative power of the verbal message is more clearly a theoretical possibility, more clearly something not present in the world of the comprehender and thus more clearly something that we can understand only if we are willing to change ourselves, if only temporarily.

There is always pressure to make a virtue out of necessity, and no doubt there are many potential educational benefits in the current communication revolution, especially in the ability of television to bring the world so close. But the ability to change the world by turning a knob can readily appear more akin to the wondrous transformations of magic than to the constructive, deliberate transformations of literacy. While the ludic assimilative power that we exercise in viewing television does enforce our sense of personal freedom, it can also readily become merely an extension of our freedom to consume, that is, an extension of a personal freedom that can express itself only in our selecting one program or one channel rather than another. Such a freedom lacks one thing—the ability to construct what is not already there.

A decline in literacy may thus be part of a general cultural movement that works to legitimatize the power of existing institutions by combining the technological domination of nature with the illusion of maximum personal freedom. This process has a clear political danger, since the freedom we are given lacks a practical and hence a political dimension. It is not a freedom to create or to destroy, or even to choose and then to act in accordance with our decisions; it is only a freedom to select some product or some idea or some opinion instead of something else, even as we are amazed at the breadth of the offerings and unable to imagine that anything has been left out.

To use the metaphor of housebuilding developed in chapter 3, we face the danger of becoming more adept at selecting patterns and colors but less adept at building the house ourselves—and even housebuilders themselves may be facing the same problem as the expansion of technology forces them to turn to a series of subcontractors, each of whom knows only how to perform one component of the total project. The increasing technological control of nature, in other words, raises the possibility that there will be an ever-increasing accumulation of symbolic knowledge and an ever-increasing inability of people at large to avail themselves of this knowledge as its principles of application become further removed from common practice. The danger we face, quite simply, is in our coming to live entirely in a world of specialists.

The plight of literacy is thus tied to the same historical process that led to its initiation, the process that Max Weber called the "disenchantment of the world." Literacy only flourishes with the decline of magical practices, yet the very decline of one form of magic creates the possibility of another form—the belief in the total technological domination of nature. Weber had the great insight that the increase in a group's collective assimilative power is not directly related to individual levels of understanding: "The increasing intellectualization and rationalization do not, therefore, indicate an increased and general knowledge of the conditions under which one lives. It means something else, namely, the knowledge or belief that if one but wished one *could* learn it at any time" (1958b:139).

There is, in other words, a profound difference between individual and collective powers of assimilation. As the gap widens between what the individual can do as part of synecdochic practice and what is possible through the application of symbolic knowledge, the range of praxis narrows. Technology seems less under the control of any one person or group and more independent, seeming to act in pursuit of its program of ever greater mastery of nature. In the triumph of ends over means, the whole purpose of praxis as self-directed activity aimed at transforming the world collapses, since activity itself becomes motivated not by individuals but by the needs of technology itself—except to the extent that we mistake the needs of technology for our own personal needs and in so doing internalize completely the synecdochic world in response to our sense of the ever-growing accommodative pressures on us to fit into the existing system.

Marshall McLuhan was correct in observing that technological changes in communication can re-create earlier, tribal forms of existence. In the new global village, as in the earlier tribe, accommodation and assimilation may well be maintained at a level of playful immediacy. In both cases there may appear to be few constraints on either the ease with which the world can be changed or the range of options open to the individual—the inhabitants of the tribal village can see no alternatives to their accommodative practices, and the inhabitants of the global village can see no limits on theirs. For too many of us today, the world of the possible is defined by the seemingly infinite variety of what already exists; like the tribal inhabitants, we thus often lack any sense of the contingency of the social world—instead it seems to dominate us as if it were an inexorable natural force. Only in the challenge to this world do we sense contingency; only in praxis, including the acts of reading and writing texts, do we most seriously threaten this world, not by changing that which is inherently malleable and thus seemingly impervious to real change, but by objectifying new modes of being that, in the very act of construction and apprehension, reveal to us the inherent possibility for change.

III. In *Orality and Literacy*, Walter Ong analyzes the relationship between literacy and technology from the perspective of the critical, "anti-literate," antitechnological tradition that lies at the center of literacy itself. It is a tradition that sees literacy and technology as working together in the Western onslaught on the indigenous practices that are rooted in the sanctity and intimacy of orality. The spoken word, animated by breath, connects us to a more spiritual, communal form of existence: "By removing words from the world of sound where they had their first origin in active human interchange and relegating them definitively to visual surface, and by otherwise exploiting visual space for the management of knowledge, print encouraged human beings to think of their own interior conscious and unconscious resources as more and more thing-like, impersonal and religiously neutral" (1982:131–32). Printing, according to Ong, "embedded the word itself deeply in the manufacturing process and made it into a kind of commodity" (p. 118). In such a conception of history, literacy is associated with untrammeled technological growth and thus with science, the intellect, and reason itself, as opposed to the forces of preservation, art, the emotions, and the imagination.

While this eschatological view of history seems to combat the positivist's claim that science and reason have dethroned religion and philosophy, in the end it really endorses the most radical claim of positivism—the separation of reason and subjective experience—while merely reversing the central argument. Reason remains aligned with logic, science, and technology and thus opposes feeling, imagination, and self-expression. Both in positivism itself and in positivist-inspired critiques of positivism, the defense of reason is associated with the domination of technology and science.

In one sense, there is good reason to connect literacy with that aspect of reason concerned with the pursuit of knowledge for its own sake and the subsequent domination of nature. The reading and writing of texts are necessarily symbolic activities that take place outside the realm of ordinary social discourse and thus apart from the strong emotional attachments that are an unavoidable part of indigenous practice. What we comprehend or create in texts must in some sense seem less immediate and less emotional than that which forms the basis of conversation; literate communication is fundamentally theoretical, even if it is richly detailed and highly personal.

We err, however, in allowing the legacy of positivism to lead us to associate theoretical activity with either the instrumental reasoning of technology or the abstract reasoning of logic. Piagetian studies of cognitive development often support such an approach by associating intelligence and formal operations with the mastery of the rules of logic. But the crucial element for literacy and for symbolic representation as a whole is not the mastery of specific rules of logic but the more general ability to judge a particular set of circumstances according to rules that are not already embedded in social practice. Internal adherence to certain rules of logic is only one of any number of possible tests of symbolic truth. What finally makes symbolic meaning theoretical is not its basis in formal logic but our recognition of its inherent contingency—our recognition, in other words, that it cannot be accepted or rejected as true simply on the basis of its correspondence with prevailing attitudes. The source for symbolic truth must be sought elsewhere—and only the prevalence of positivist attitudes leads us to locate this "elsewhere" in the methodology of the natural sciences or formal logic and not, for example, in the universality of aesthetic experience or in the imaginative power of historical understanding.

The essential dichotomy underlying literacy is not, as either Ong or his positivist opponents would have it, between thinking and feeling, science and art, modernism and tradition, or writing and speaking. Rather, it is between those forms of practice that tend to reify the existing social order and those forms of praxis, including the acts of creating and comprehending texts, that tend to expose its contingent nature. Only for the positivist or for those "opponents" of positivism who have made their accommodation with it is reason blind adherence to the pursuit of logic or science without regard to human goals. For those who seek neither a truth of the mind nor a truth of the heart, that is, for those who seek only a unifying human truth, reason is something quite different—it is instead the active working of the self, directed by understanding, in pursuit of freely chosen goals. Literacy, as the instrument of reason, is thus neither the servant of technology nor the enemy of untainted emotion.

The future of literacy thus depends upon the maintenance of a tension between the uses of language that make us more able and less able to recognize the continual, inherent possibility for us to change ourselves as a means of changing the world. In the absence of literacy, we are far more likely to be unable to distinguish the historical and the natural world: "The pressures and limits of what can ultimately be seen as a specific economic, political, and cultural system," writes Raymond Williams in interpreting Gramsci, "seem to most of us the pressures and limits of simple experience and common sense" (1977:110). The practical activity of reading and writing, as much as anything else, pushes back those limits and does so regardless of the actual political content of the text. While some uses of reading and writing may tend to harden our sense of the limits of the possible, such ritualistic acts must not be confused with literacy. Any literate act, even when it seeks to justify the existing social order and thus argues against our effecting any change, nevertheless still reveals something important about the possibility for change implicit in our condition. Acts of preservation are grounded in the same principles of literacy as acts of revolution, for both depend upon symbolic transformations to reveal the possibility for human action.

Gadamer recognized, as borne out by many contemporary social movements, that "preservation is as much a freely-chosen action as revolution and renewal" (1982:250). Indeed, it is Gadamer's lasting contribution, embodied in the title of his great work *Truth and Method*, that the truth we seek in order to live can never be equated with knowledge reached through the application of any one method. Reason is not some abstract force or method inevitably at odds with the prevailing beliefs as expressed in prejudgments and traditions; it is instead the continual activity of individuals, themselves the products of history, aimed at seeking understanding in the acts of objectification that distance themselves, without ever freeing them totally, from their own historical condition.

The life of reason that forms the basis of literacy therefore lies neither in the continued domination of the past nor in our total liberation from it but only in the dynamic interaction between what is and what can be. Literacy as such is neither simply a reactionary force nor a radical force, since it constantly resists the pull of the right to legitimatize the status quo unconditionally and the pull of the left to liberate us totally from it. It simultaneously resists both the reactionary attraction of a golden age that was initially destroyed and is continually being destroyed by symbolic activity and the radical attraction of a heavenly city that will liberate individuals from all forms of synecdochic practice and belief.

The domination of either the utopias of the left or the ideologies of the right represents an equally serious threat to literacy, since each represents a vision of society that fails to recognize the fundamental need to maintain a sense of tension between the present and both the past and the future. Theoretical views of the past and the future are themselves often the products of literacy, and both are needed, again in Ricoeur's words, to "stand out against a common background of non-congruence (behind or ahead) in relation to a concept of reality which is revealed only in effective practice" (1981a:240). Both ideologies and utopias, in other words, posit ideal worlds that illuminate the real one, and as such both must exist not in practice but only as theoretical, literate projections of ways of being in the world. Together, writes Ricoeur, they represent necessary "gaps or discordances in relation to the real course of things": "A social group without ideology and utopia would be without a plan, without a distance from itself, without a self-representation. It would be a society without a global project, consigned to a history fragmented into events which are all equal and insignificant" (p. 241).

Is it really possible, we might ask, for either ideological or utopian thinking to disappear or to dominate social experience so completely that there is little motivating force for the praxis that consummates in literacy? The answer to this crucial question is that it seems to be not only possible but quite natural. For long stretches of time, various people have possessed the technology for coding language and have found few if any literate uses for this technology. Indeed, it may well be that the near-total domination of consciousness by either ideologies or utopias, or both in alternation, constitutes the norm of human existence, with periods of active struggle involving the reading and writing of texts as singular exceptions. The great explosion in literacy in the last century and a half may well have been the result of an inescapable conflict between existing patterns of belief and practice precipitated by a worldwide revolution in economic production, and it is possible that certain alterations to this mode of industrial production, whatever their benefits in various areas, may nevertheless represent a challenge to literacy itself.

Neither a technological, liberated future nor a traditional, idyllic past in itself represents the ultimate threat to literacy, but only the total domination of one at the expense of the other; that is, what finally threatens the future of literacy is only the uncontested triumph of the present. Were the dialogue between past and present to collapse, we would experience the illusion of living in a world that had freed us from having to make the distinction between what we can desire and what we can possess—but this apparent freedom would likely come from the constriction, not the growth, of our power to act constructively. A world capable of seemingly endless, magical transformations requires little accommodation and thus is capable of appearing to us as progressive and liberating even if it is ruthlessly totalitarian. In such a world, *reason* either will be trumpeted as an omnipotent god providing for our technological domination of nature or will be condemned as an archetypal malcontent intruding on our world of personal freedom; meanwhile, as the world becomes steadily more authoritarian, popular attention generally will likely focus on *authority* as the greatest threat to human fulfillment. With the total triumph of the present, any positive sense of authority as a protection against consensus will have to be rejected as an unacceptable bridle on individual liberty. It will be construed as an entirely negative force—one attempting to dominate and even enslave individuals by restricting their range of choice.

Only the maintenance of an image of an ideal past or an ideal future, that is, only the continued existence (but not domination) of ideologies and utopias as theoretical models of a better world, allows us to see authority, not as an inhibitor, but as a positive force predisposing us to suspend the temptation to act immediately and leading us instead to weigh our decisions in the light of what synecdochic practice informs us is a source of symbolic understanding. Without an ability to see ideologies and utopias as theoretical possibilities, we cannot understand Gadamer's claim that authority has "nothing to do with obedience, but rather with knowledge" (1982:248); without an ability to reject the present, we cannot see authority as a continual reminder that there are ways of being in the world that we can understand only through acts of interpretation requiring our reliance upon the guidance and the expertise of others with whom we have no synecdochic connection. As Richard Sennett remarks, belief in authority ultimately requires us to forsake what is immediate for what, in some crucial sense, can never exist; such a belief, he concludes, is "inherently an act of imagination" (1980:197).

For the present to triumph finally, authority must be discounted as a force that seeks to constrain choice and thus to rob us of our freedom by encouraging us to follow one path as better than another. The value of authority must be discounted if we are to preserve the illusion that we live in a world that no longer requires accommodation in any area, including reading. In this sense, the threat to literacy inherent in the triumph of the present is evident in the aspect of contemporary literary criticism that is concerned largely with liberating reading by unmasking the ideological foundation of all authority. The discrediting of certain authorities can clearly play a crucial role in freeing us from the domination of ideology—we must learn that any inherited, idealized view of the world is distorted by power relationships. Contemporary literary critics, however, often seem eager to equate the historical limitations of any one authority with a rejection of authorities in general. In so doing, they are apt to reject the tyranny of the past, as expressed in ideologies, for the illusion of a utopian future characterized by total, unrestrained communication between equals. There are no final authorities, we are told, only "interpretive communities," and in a limited sense such a statement is irrefutable.

We err, and in so doing threaten literacy, when we fail to recognize the crucial distinction between "interpretive communities" to which we belong on the basis of immediate social relations and those to which we belong by virtue of our efforts to transcend the limits imposed by those relations. While classifying all judgments as "subjective," as Bleich does, creates the illusion of liberating reading from the tyranny of synecdochic pressure, its effect is just the opposite. Inasmuch as we cannot rely on authorities whose judgments are recognized and accepted even by those with different practices, we are less likely to be able to affirm symbolic meanings in the face of the unavoidable personal pressures created by synecdochic ties. Only those critics who are themselves completely a part of an infinitely malleable present could be so fearful of the domination of the clear voices of authorities with whom we have no social contact while remaining unsuspecting of the domination of the whispered voices of those with whom we live.

To create and comprehend texts is always a difficult act, almost a disloyal one, for we are required to forsake what we share with others. The possibility of literacy, therefore, rests in part on the belief in the higher authority of some distant, nonpersonal other. While the total domination of this other may seem to be a violation of one of our most cherished democratic impulses (that individual opinion, no matter how humble, does have its own sanctity), its total abandonment poses an equally great, and more likely, threat—the death of the past and the future as living images of other ways of being. In this sense, E. D. Hirsch is correct in seeing "formalism" and "pluralism" as dual threats—educational practice is often too intent upon doing little more than giving everyone coding skills and the freedom to do with them as they please.

Yet Hirsch's interest in restoring authority is finally as limited as the efforts of those trying to overthrow it. To assume that, unless authority can somehow escape the limits of our historical condition, it must somehow be rejected as merely opinion is to accept wholeheartedly and uncritically the fundamental premise of positivism. We do not seek objective truth in order to escape error or relativism—there can be no denying Wittgenstein's insight that "at the foundation of well-founded belief lies belief that is not founded" (1969:33e). Positivistic approaches in general misconstrue our basic human interest in obtaining truth free of opinion. We seek such truth, not as a means of sustaining the illusion that we have the power of the scientist, but, as Zygmunt Bauman writes, as a "safeguard against imposed consensus, as an appeal against unacceptable outcome of communal negotiation. ... The urge for objective understanding as distinct from simply communal consensus is generated by restrictions imposed upon equality and democracy by the structure of domination which underlies the negotiating process" (1978:224). Our living with others both establishes the need for and sets the limits to the objectivity of authority and in so doing prevents us from ever totally freeing ourselves from a structure of domination, despite our constant and necessary utopian aspirations.

To believe, as Hirsch suggests, that all acts of interpretation, including reading, can be governed by a prescribed set of authorities is to fall prey to the demagogic ideology of the political arena, yet to see all such acts of interpretation as a struggle between differing communities, each sustained by a sense of its own legitimacy, is to fall prey to what Hirsch senses to be an even more elusive, hence more beguiling, danger, the utopianism of the marketplace. It is to believe that a world of free and open communication between equals can exist in practice as readily as it can in theory—if only people were more open-minded. It is to believe that in the triumph of the present we will all live in a totally tolerant world, one able to accept, or at least able to explain, all human activity as an acceptable form of accommodation.

In such a world, however, there will be no praxis, since anything that we do, no matter how extraordinary, anomalous, or even outrageous, will be readily accepted as an acceptable form of idiosyncratic practice. "What is" is what will be and what can be! Such a world will be capable of finding endless uses for written language but few uses for literacy. It will be a tolerant, sentimental world but one that cannot readily distinguish between that which merely reinforces the existing social order and that which threatens it, and thus society will hardly know where to begin transforming itself. It will be a world in which many of our most pressing social problems will disappear, not because they are solved, but because they have become part of the nature of things. This new world will view the death of literacy, not as its disappearance, but as its triumph: all the people of the world will be able to encode and decode oral language, but there will no longer be a separate world of texts. There will be people reading and writing books, and there will be bountiful interpretations of them, all variations of a malleable present, but there will be no authorities, no critical tradition struggling to define the boundaries between past, present, and future. The overthrow of authority, like the end of literacy it accompanies, will be celebrated as the liberation of humanity, but in this liberation from authority lies the origins of the most pernicious forms of authoritarianism—the total domination of existing social practice.

What will prevent such a future for literacy and what will thus ensure its survival forever is precisely what has nourished it in the past—namely, the practical activities of people intent upon understanding their past and shaping their future. No matter how powerful the technology of the future may grow, the world will likely never be malleable enough to please everyone. Those who are dissatisfied with the world will continue to try to change it, and literacy will exist in their use of the language of the present to articulate a personal sense of their past and their future. The force that directs their action will be *reason*, and that which gives value to it will be *authority*; the activity itself will be *praxis*, and the language of praxis will be *literacy*. Reason and authority will continue to be the protectors of literacy, just as praxis will be its expression. In the final analysis, the fate of literacy lies not with the mechanics of transcribing language or even with pedagogic and curriculum reforms in language education, nor does it lie with technological revolutions in communication, with satellites, computers, and word processors—it lies instead only with the continued efforts of people everywhere to objectify in language that which does not yet exist. Like the solitary protagonist of *Pilgrim's Progress*, the literate among us, now and in the future, will always bear a great burden as they stand with a book in their hands and their backs to their ravaged houses, reading and wondering aloud, "What shall I do?"

Works Cited

Abrams, Meyer. 1953. *The Mirror and the Lamp: Romantic Theory and the Critical Tradition.* New York: Oxford University Press.

Adler, Sol. 1979. *Poverty, Children, and Their Language: Implications for Teaching and Treating.* New York: Grune and Stratton.

Altick, Richard. 1957. *The English Common Reader: A Social History of the Mass Reading Public, 1800–1900.* Chicago: University of Chicago Press.

Applebee, Arthur. 1974. *Tradition and Reform in the Teaching of English.* Urbana, Ill.: National Council of Teachers of English.

Applebee, Arthur, with Anne Austen and Fran Lehr. 1981. *Writing in the Secondary School: English and the Content Areas.* Urbana, Ill.: National Council of Teachers of English.

Arnold, Matthew. [1868] 1961. *Culture and Anarchy.* Reprint, ed. J. Dover Wilson. Cambridge: Cambridge University Press.

Bates, Elizabeth. 1976. *Language and Context: The Acquisition of Pragmatics.* New York: Academic Press.

Bauman, Zygmunt. 1973. *Culture as Praxis.* Boston: Routledge and Kegan Paul.

———. 1978. *Hermeneutics and Social Science.* New York: Columbia University Press.

Bell, Terrel H. 1982. "Ask Them Yourself." *Family Weekly* 6 (June):2.

Bernstein, Basil. 1967. "The Role of Speech in the Development and the Transmission of Culture." In *Perspectives on Learning,* ed. C. L. Klept and W. A. Hohman, 15–45. N.p.: Mental Materials Center.

———. 1977. *Class, Codes, and Control.* Vol. 3, *Towards a Theory of Educational Transmissions.* Rev. ed. London: Routledge and Kegan Paul.

———. 1981. "Codes, Modalities, and the Process of Cultural Reproduction: A Model." *Language in Society* 10:327–63.

Bleich, David. 1975. "The Subjective Character of Critical Interpretation." *College English* 36:739–55.

———. 1976. "The Subjective Paradigm in Science, Psychology, and Criticism." *New Literary History* 7:313–34.

Bloomfield, Leonard. 1933. *Language.* New York: Holt.

Bormouth, John. 1978. "Literacy Policy, Reality, and Writing Instruction." In *Perspectives on Literacy: Proceedings of the 1977 Perspectives on Literacy Conference,* ed. R. Beach and P. David Pearson, 13–41. Minneapolis: University of Minnesota, College of Education.

Bourdieu, Pierre, and Jean-Claude Passeron. 1977. *Reproduction: In Education, Society, and Culture.* Trans. Richard Nice. Beverly Hills, Calif.: Sage Publications.

Bowles, Samuel, and Herbert Gintis. 1976. *Schooling in Capitalist America: Education Reform and the Contradiction of Economic Life.* New York: Basic Books.

Britton, James. [1970] 1972. *Language and Learning.* Reprint. Hammondsworth: Pelican-Penguin.

Bruner, Jerome. 1965. "The Growth of Mind." *American Psychologist* 20:1007–17.

———. 1973. "The Perfectability of Intellect." In *The Relevance of Education,* ed. Anita Gil, 3–19. New York: Norton.

———. 1975. "From Communication to Language—a Psychological Perspective." *Cognition* 3:255–87.

Bruner, Jerome, with Patricia Greenfield. 1966. "An Overview." In *Studies in Cognitive Growth,* ed. J. Bruner, Rose R. Olver, and Patricia M. Greenfield, 319–26. New York: Wiley.

———. 1973. "Culture and Cognitive Growth." In *The Relevance of Education,* ed. Anita Gil, 20–51. New York: Norton.

Burgess, Mary Ayres. 1921. *The Measurement of Silent Reading.* New York: Sage Foundation.

Burke, Kenneth. [1945] 1969. *A Grammar of Motives.* Reprint. Berkeley: University of California Press.

———. 1955. *A Rhetoric of Motives.* New York: Braziller.

———. 1966. *Language as Symbolic Action: Essays on Life, Literature, and Method.* Berkeley: University of California Press.

———. 1984. *Permanence and Change: An Anatomy of Purpose.* 3d ed. Berkeley: University of California Press.

Calhoun, Daniel. 1973. *The Intelligence of a People.* Princeton: Princeton University Press.

Carothers, J. C. 1959. "Culture, Psychiatry, and the Written Word." *Psychiatry* 22:307–20.

Carpenter, George R., Franklin T. Baker, and Fred N. Scott. 1913. *The Teaching of English in the Elementary and Secondary School.* New York: Longmans, Green.

Carroll, John B., and Jeanne Chall, eds. 1975. *Toward a Literate Society.* New York: McGraw-Hill.

Cassirer, Ernst. 1968. *The Philosophy of Symbolic Forms. Vol. 1, Language.* Trans. Ralph Mannheim. New Haven, Yale University Press.

Channing, Edward T. 1856. *Lectures Read to the Seniors in Harvard College.* Boston: Ticknor and Fields.

Chomsky, Noam. 1979. *Language and Responsibility.* Trans. John Viertel. New York: Pantheon.

———. 1980. *Rules and Representations.* New York: Columbia University Press.

Cicourel, Aaron. 1973. *Cognitive Sociology.* New York: Free Press.

Cipolla, Carlo. 1980. *Literacy and Development in the West.* Baltimore: Penguin Books.

Cole, Michael, and Sylvia Scribner. 1974. *Culture and Thought: A Psychological Introduction.* New York: Wiley.

Collingwood, R. G. 1956. *The Idea of History.* London: Oxford University Press.

Collins, John Churton. 1887. "Can English Be Taught?" *Nineteenth Century* 22:642–58.

Collins, Randall. 1977. "Some Comparative Principles of Educational Stratification." *Harvard Education Review* 47:1–27.

DeLaguna, Grace. [1927] 1970. *Speech: Its Function and Development.* Reprint. College Park, Md.: McGrath.

Diamond, Stanley. 1963. "The Search for the Primitive." In *Man's Image in Medicine and Anthropology,* ed. Iago Galdston, 62–115. New York: International Universities Press for the Institute of Social and Historical Medicine.

Diehl, William. 1979. "The Variable and Symbolic Nature of Functional Literacy: A Historical Review and Critique of Research." M.A. thesis, Indiana University. ERIC Document 186-868.

Dillon, George. 1981. *Constructing Texts: Elements of a Theory of Composition and Style.* Bloomington: Indiana University Press.

Donoghue, Dennis. 1981. *Ferocious Alphabets.* Boston: Little, Brown.

Douglas, Wallace. 1976. "Rhetoric for the Meritocracy." In *English in America: A Radical View of the Profession,* ed. Richard Ohmann, 97–132. New York: Oxford University Press.

Durkheim, Emile. 1933. *The Division of Labour in Society.* Trans. G. Simpson. Glencoe, Ill.: Free Press.

Edwards, John R. 1979. *Language and Disadvantage.* New York: Elsevier.

Ekvall, R. B. 1964. *Religious Observance in Tibet.* Chicago: University of Chicago Press.

Eliot, Charles W. 1909. *Educational Reform: Essays and Addresses.* New York: Century.

Ellul, Jacques. 1980. *The Technological System.* Trans. Joachim Neugroschel. New York: Continuum.

Emig, Janet. 1983. "Freedom and Literacy." In *The Web of Meaning,* ed. D. Goswami and Maureen Butler, 171–78. Upper Montclair, N.J.: Boynton/Cook.

Entwistle, Harold. 1978. *Class, Culture, and Education.* London: Methuen.

Fadiman, Clifton, and James Howard. 1979. *Empty Pages: A Search for Writing Competence in School and Society.* Belmont, Calif.: Fearon Pitman.

Femia, Joseph. 1981. *Gramsci's Political Thought: Hegemony, Consciousness, and the Revolutionary Process.* Oxford: Clarendon Press.

Fish, Stanley. 1980. "Interpreting the 'Variorum.'" In *Reader-Response Criticism: From Formalism to Post-Structuralism,* ed. Jane P. Tomkins, 164–84. Baltimore: Johns Hopkins University Press.

Folger, John K., and Charles B. Nam. 1967. *Education of the American Population: A 1960 Census Monograph.* Washington, D.C.: Government Printing Office.

Freud, Sigmund. 1960. *Jokes and Their Relation to the Unconscious.* Trans. James Strachey. New York: Norton.

Fries, Charles. 1962. *Linguistics and Reading.* New York: Holt, Rinehart and Winston.

Gadamer, Hans-Georg. 1981. *Reason in the Age of Science.* Trans. Frederick G. Lawrence. Cambridge, Mass.: MIT Press.

———. 1982. *Truth and Method.* New York: Crossroad.

Gay, John, and Michael Cole. 1967. *The New Mathematics and an Old Culture.* New York: Holt, Rinehart and Winston.

Genovese, Eugene. 1974. *Roll, Jordan, Roll: The World the Slaves Made.* New York: Random House, Pantheon Books.

Gleitman, Henry, and Lila Gleitman. 1979. "Language Use and Language Judgment." In *Individual Differences in Language Ability and Language Behavior,* ed. Charles J. Fillmore, Daniel Kempler, and William Wang, 103–26. New York: Academic Press.

Goody, Jack. 1968a. "Introduction." In *Literacy in Traditional Societies,* ed. Goody, 1–26. Cambridge: Cambridge University Press.

———. 1968b. "Restricted Literacy in Northern Ghana." In *Literacy in Traditional Societies,* ed. Goody, 199–264. Cambridge: Cambridge University Press.

———. 1977. *The Domestication of the Savage Mind.* Cambridge: Cambridge University Press.

Goody, Jack, and Ian Watt. 1963. "The Consequences of Literacy." *Comparative Studies in Society and History* 3:304–45.

Graff, Gerald. 1979. *Literature against Itself: Literary Ideas in Modern Society.* Chicago: University of Chicago Press.

Graff, Harvey. 1979. *The Literacy Myth: Literacy and Social Structure in the Nineteenth Century City.* New York: Academic Press.

Gramsci, Antonio. 1971. *Selections from the Prison Notebooks,* ed. Quintin Hoare and Geoffrey Newell Smith. New York: International Publishers.

Grandgent, Charles H. 1930. "The Modern Languages." In *The Development of Harvard University since the Inauguration of President Eliot, 1869–1929,* ed. Samuel E. Morison, 65–81. Cambridge: Harvard University Press.

Graves, Donald H. 1978. *Balance the Basics: Let Them Write.* New York: Ford Foundation.

Greenfield, Patricia M. 1972. "Oral or Written Language: The Consequences for Cognitive Development in Africa, the United States, and England." *Language and Speech* 15:169–78.

Greenfield, Patricia, and Patricia Zukow. 1978. "Why Do Children Say What They Say When They Say It? An Experimental Approach to the Psychogenesis of Presupposition." In *Children's Language,* vol. 1, ed. Keith Nelson, 287–336. New York: Gardner Press.

Grice, H. P. 1975. "Logic and Conversation." In *Syntax and Semantics,* vol. 3, *Speech Acts,* ed. Peter Cole and Jerry Morgan, 41–58. New York: Academic Press.

Gudschinsky, Sarah. 1976. *Literacy: The Growing Influence of Linguistics.* The Hague: Mouton.

Habermas, Jurgen. 1983. *Theory and Practice.* Trans. John Viertel. Boston: Beacon Press.

Hallpike, Christopher Robin. 1979. *The Foundations of Primitive Thought.* New York: Oxford University Press.

Havelock, Eric. 1963. *Preface to Plato.* Cambridge, Mass.: Harvard University Press, Belknap Press.

———. 1971. "Prologue to Greek Literacy." Lecture in Memory of Louise Taft Semple. 2d ser. University of Cincinnati.

———. 1976. *Origins of Western Literacy.* Toronto: Ontario Institute for Studies in Education.

Heidegger, Martin. 1978. *Being and Time.* Trans. John Macquarrie and Edward Robinson. Oxford: Basil Blackwell.

Henson, Josiah. [1849] 1970. *Father Henson's Story of His Own Life.* Reprint. Upper Saddle River, N.J.: Literature House.

Hirsch, E. D. 1983. "Cultural Literacy." *American Scholar* 52:159–69.

Hoggart, Richard. 1961. *The Uses of Literacy: Changing Patterns in English Mass Culture.* Boston: Beacon Press.

———. 1982. *An English Temper: Essays in Education, Culture, and Communication.* New York: Oxford University Press.

Hook, J. N. 1979. *A Long Way Together: A Personal View of NCTE's First Sixty-Seven Years.* Urbana, Ill.: National Council of Teachers of English.

Horton, Robin. 1970. "African Thought and Western Science." In *Rationality,* ed. Bryan Wilson, 131–71. New York: Harper and Row.

Hosic, James Fleming. 1917. *The Reorganization of English in the Secondary Schools.* Bureau of Education Bulletin, no. 2. Washington, D.C.: Government Printing Office.

Huey, Edmund Burke. [1908] 1968. *The Psychology and Pedagogy of Reading.* Reprint. Cambridge: MIT Press.

Huizinga, Johan. 1955. *Homo Ludens: A Study of the Play-Element in Culture.* Boston: Beacon.

Hunter, Carman St. John, with David Harman. 1979. *Adult Illiteracy in the United States: A Report to the Ford Foundation.* New York: McGraw-Hill.

Hutchins, Edwin. 1980. *Culture and Inference: A Trobriand Case Study.* Cambridge, Mass.: Harvard University Press.

Iser, Wolfgang. 1974. *The Reading Process: A Phenomenological Approach.* Baltimore: Johns Hopkins University Press.

Jameson, Fredric. 1972. *The Prison-House of Language: A Critical Account of Structuralism and Russian Formalism.* Princeton: Princeton University Press.

———. 1981. *The Political Unconscious: Narrative as a Socially Symbolic Act.* Ithaca: Cornell University Press.

Jenkins, J., and A. Lieberman. 1972. "Background to the Conference." In *Language by Ear and by Eye,* ed. James Kavanaugh and Ignatius Mattingly, 1–2. Cambridge, Mass.: MIT Press.

Johansson, Egil. 1981. "The History of Literacy in Sweden." In *Literacy and Social Development in the West: A Reader,* ed. Harvey Graff, 151–82. Cambridge: Cambridge University Press.

Kay-Shuttleworth, James. 1970. "James Kay-Shuttleworth and Pupil Teachers." In *Nineteenth Century Education,* ed. Eric Midwinter, 81. New York: Harper and Row.

Keller, Helen. 1903. *The Story of My Life.* New York: Doubleday, Page.

Kelly, Louis. 1969. *Twenty-Five Centuries of Language Teaching: An Inquiry into the Science, Art, and Development of Language Teaching Methodology, 500 B.C.–1969.* Rowley, Mass.: Newbury House.

Kilminster, Richard. 1979. *Praxis and Method: A Sociological Dialogue with Lukács, Gramsci, and the Early Frankfurt School.* Boston: Routledge and Kegan Paul.

Kohlberg, Lawrence. 1970. "The Child as a Moral Philosopher." In *Readings in Developmental Psychology Today,* ed. P. Cramer, 109–15. Del Mar, Calif.: CRM.

Kozol, Jonathan. 1985. *Illiterate America.* Garden City: Doubleday, Anchor Press.

Kroll, Barry, and Roberta Vann. 1981. Introduction to *Exploring Speaking-Writing Relationships: Connections and Contrasts,* vii–xi. Urbana, Ill.: National Council of Teachers of English.

Labov, William. 1973. "The Logic of Nonstandard English." In *The Politics of Literature: Dissenting Essays on the Teaching of English,* ed. Louis Kampf and Paul Lauter, 194–244. New York: Vintage Books.

Leavis, F. R., and Denys Thompson. 1942. *Culture and Environment.* London: Chatto and Windus.

Lévi-Strauss, Claude. 1966. *The Savage Mind.* Chicago: University of Chicago Press.

———. 1974. *Tristes Tropiques.* Trans. John and Doreen Weightman. New York: Atheneum.

Levitas, Maurice. 1974. *Marxist Perspectives in the Sociology of Education.* Boston: Routledge and Kegan Paul.

Lienhardt, Godfrey. 1961. *Divinity and Experience: The Religion of the Dinka.* Oxford: Clarendon Press.

Loban, Walter. 1978. "Relationship between Language and Literacy." In *Perspectives on Literacy: Proceedings of the 1977 Perspectives on Literacy Conference,* ed. R. Beach and P. David Pearson, 97–109. Minneapolis: University of Minnesota, College of Education.

Lockridge, Kenneth. 1974. *Literacy in Colonial New England: An Inquiry into the Social Context of Literacy in the Early Modern West.* New York: Norton.

Lukács, Georg. 1971. *History and Class Consciousness: Studies in Marxist Dialectics.* Trans. Rodney Livingstone. Cambridge, Mass.: MIT Press.

Luria, Aleksandr Romanovich. 1976. *Cognitive Development: Its Cultural and Social Foundations,* ed. Michael Cole. Trans. Martin Lopez-Morillas and Lynn Solotaroff. Cambridge, Mass.: Harvard University Press.

McGuffey's Sixth Eclectic Reader. [1879] 1962. Reprint. New York: New American Library, Signet Classics.

McLuhan, H. Marshall. 1962. *The Gutenberg Galaxy: The Making of Typographic Man.* Toronto: University of Toronto Press.

———. 1964. *Understanding Media: The Extensions of Man.* New York: McGraw-Hill.

Malinowski, Bronislaw. 1923. "The Problem of Meaning in Primitive Languages." In *The Meaning of Meaning: A Study of the Influence of Language upon Thought and of the Science of Symbolism,* ed. C. K. Ogden and I. A. Richards, 296–336. London: Routledge and Kegan Paul.

Marx, Karl. 1967a. *Capital.* 3 vols, ed. Frederick Engels. Trans. Samuel Moore and Edward Aveling. New York: International Publishers.

———. 1967b. *The Writings of the Young Marx on Philosophy and Society,* ed. Lloyd Easton and Kurt Guddat. New York: Doubleday, Anchor Press.

Mathews, Mitford M. 1966. *Teaching to Read, Historically Considered.* Chicago: University of Chicago Press.

Mathieson, Margaret. 1975. *The Preachers of Culture: A Study of English and Its Teachers.* London: Allen and Unwin.

Mead, George Herbert. [1934] 1967. *Mind, Self, and Society: From the Standpoint of a Social Behaviorist.* Reprint, ed. Charles W. Morris. Chicago: University of Chicago Press.

Mead, Margaret. 1943. "Our Educational Emphases in Primitive Perspective." *American Journal of Sociology* 48:633–39.

Mill, John Stuart. 1924. *The Autobiography of John Stuart Mill.* New York: Columbia University Press.

Mitchell-Kernan, Claudia. 1980. Introduction to *Runnin' Down Some Lines: Language and Culture of Black Teenagers,* by Edith Folb. Cambridge, Mass.: Harvard University Press.

National Assessment of Educational Progress. 1981. *Reading, Thinking, and Writing: Results from the 1979–80 National Assessment of Reading and Literature.* Denver: Education Commission of the States.

National Council of Teachers of English. 1979. *Councilgram* 41 (May).

Needham, Rodney. 1972. *Belief, Language, and Experience.* Oxford: Basil Blackwell.

Ohmann, Richard. 1976. *English in America: A Radical View of the Profession.* New York: Oxford University Press.

Olson, David, and Nancy Nickerson. 1977. "The Contexts of Comprehension: On Children's Understanding of the Relations between Active and Passive Sentences." *Journal of Experimental Child Psychology* 23:402–14.

———. 1978. "Language Development through the School Years: Learning to Confine the Interpretation to the Information in the Text." In *Children's Language,* vol. 1, ed. Keith Nelson, 117–69. New York: Gardner Press.

O'Neil, Wayne. 1970. "Properly Literate." *Harvard Education Review* 40:260–63.

Ong, Walter. 1971. *Rhetoric, Romance, and Technology.* Ithaca: Cornell University Press.

———. 1977. *Interfaces of the Word.* Ithaca: Cornell University Press.

———. 1982. *Orality and Literacy: The Technologizing of the Word.* New York: Methuen.

Parker, William Riley. 1967. "Where Do English Departments Come From?" *College English* 28:339–51.

Peirce, Charles Sanders. 1932. *Collected Papers,* ed. C. Jartshorne and P. Weiss. Cambridge, Mass.: Harvard University Press.

Phelps, William Lyon. 1939. *Autobiography with Letters.* New York: Oxford University Press.

Piaget, Jean. 1954. *The Construction of Reality in the Child.* Trans. Margaret Cook. New York: Basic Books.

————. 1962. *Play, Dreams, and Imitation in Childhood*. Trans. C. Gattegno and F. M. Hodgson. New York: Norton.

Poster, Mark. 1975. *Existential Marxism in Postwar France: From Sartre to Althusser*. Princeton: Princeton University Press.

Report of the Committee of Ten on Secondary School Studies. 1894. New York: American Book Company for the National Education Association.

Rice, Joseph Mayer. 1893. *The Public School System of the United States*. New York: Century.

Richards, I. A. 1925. *Principles of Literary Criticism*. New York: Harcourt, Brace.

————. 1929. *Practical Criticism: A Study in Literary Judgment*. New York: Harcourt, Brace and World.

Ricoeur, Paul. 1976. *Interpretation Theory: Discourse and the Surplus of Meaning*. Fort Worth: Texas Christian University Press.

————. 1978. "Creativity in Language: Word, Polysemy, Metaphor." In *The Philosophy of Paul Ricoeur*, ed. Charles E. Reagan and David Stewart, 120–33. Boston: Beacon Press.

————. 1981a. "Science and Ideology." In *Hermeneutics and the Human Sciences*, ed. John B. Thompson, 222–46. New York: Cambridge University Press.

————. 1981b. "The Task of Hermeneutics." In *Hermeneutics and the Human Sciences*, ed. John B. Thompson, 43–62. New York: Cambridge University Press.

Sapir, Edward. 1956. "Language." In *Culture, Language, and Personality*, ed. David Mandelbaum, 1–44. Berkeley: University of California Press.

Sartre, Jean-Paul. 1963. *The Search for a Method*. Trans. Hazel Barnes. New York: Knopf.

Schütze, Martin. 1933. *Academic Illusions in the Fields of Letters and the Arts*. Chicago: University of Chicago Press.

Scribner, Sylvia, and Michael Cole. 1981. *The Psychology of Literacy*. Cambridge, Mass.: Harvard University Press.

Searle, John. 1979. "Metaphor." In *Metaphor and Truth*, ed. Andrew Ortony, 92–123. Cambridge: Cambridge University Press.

Sennett, Richard. 1980. *Authority*. New York: Knopf.

Shaughnessey, Mina. 1977. *Errors and Expectations: A Guide for the Teacher of Basic Writing*. New York: Oxford University Press.

Smith, Frank. 1977. "Making Sense of Reading—and of Reading Instruction." *Harvard Education Review* 47:386–95.

Smith, Nila. 1939. *American Reading Instruction*. New York: Silver Burdett.

Soltow, Lee, and Edward Stevens. 1981. *The Rise of Literacy and the Common School in the United States: A Socioeconomic Analysis to 1870*. Chicago: University of Chicago Press.

Steiner, George. 1975. *After Babel: Aspects of Language and Translation*. New York: Oxford University Press.

Stephen, Leslie. 1887. "The Study of English Literature." *Cornhill* 55:486–508.

Stewart, Donald C. 1976. "'The Unteachable Subject': A Cautionary Tale." *Change* 8:48–51, 63.

Sticht, Thomas, ed. 1975. *Reading for Working: A Functional Literacy Anthology*. Alexandria, Va.: Human Resources Research Organization.

Sticht, Thomas, Lawrence J. Beck, Robert N. Hauke, Glenn M. Kleiman, and James H. James. *Auding and Reading: A Developmental Model.* Alexandria, Va.: Human Resources Research Organization.

Tambiah, S. J. 1968. "Literacy in a Buddhist Village in North-East Thailand." In *Literacy in Traditional Societies*, ed. Jack Goody, 86–131. Cambridge: Cambridge University Press.

———. 1973. "Form and Meaning of Magical Acts: A Point of View." In *Modes of Thought: Essays on Thinking in Western and Non-Western Societies*, ed. Robin Horton and Ruth Finneran, 199–229. London: Faber and Faber.

Tchudi [Judy], Stephen. 1980. *The ABCs of Literacy: A Guide for Parents and Educators.* New York: Oxford University Press.

Thompson, Edward P. 1963. *The Making of the English Working Class.* New York: Random House.

UNESCO. 1976. *Experimental World Literacy Program: A Critical Assessment.* Paris: UNESCO Press.

Vachek, Josef. 1973. *Written Language: General Problems and Problems of English.* The Hague: Mouton.

Vygotsky, Lev Semenovich. 1962. *Thought and Language.* Trans. Eugenia Hanfmann and Gertrude Vakar. Cambridge, Mass.: MIT Press.

———. 1978. *Mind in Society.* ed. Michael Cole, Vera John-Steiner, Sylvia Scribner, and Ellen Souberman. Cambridge, Mass.: Harvard University Press.

Weber, Max. 1958a. "The Chinese Literati." In *From Max Weber: Essays in Sociology*, ed. H. H. Gerth and C. Wright Mills, 416–44. New York: Oxford University Press.

———. 1958b. "Science as a Vocation." In *From Max Weber: Essays in Sociology*, ed. H. H. Gerth and C. Wright Mills, 129–56. New York: Oxford University Press.

Williams, Raymond. 1977. *Marxism and Literature.* New York: Oxford University Press.

Wiseman, D. J. 1962. *The Expansion of Assyrian Studies.* Inaugural Lectures, School of Oriental and African Studies. London: University of London.

Wittgenstein, Ludwig. 1969. *On Certainty*, ed. G. E. M. Anscombe and G. H. von Wright. Trans. Denis Paul and G. E. M. Anscombe. New York: Harper.

Yerkes, Robert M. 1921. *Psychological Examining in the United States Army.* Memoirs of the National Academy of Sciences, no. 15. Washington, D.C.: Government Printing Office.

Index

194